Parametric Modeling

with Autodesk Inventor® R10

Randy H. Shih

Oregon Institute of Technology

ISBN:
1-58503-237-9 (Text only)
1-58503-238-7 (Text bundled with 180-day license Inventor R 10)

SDC
PUBLICATIONS
Mission, Kansas

Schroff Development Corporation

P.O. Box 1334
Mission KS 66222
(913) 262-2664
www.schroff.com

Trademarks

The following are registered trademarks of Autodesk, Inc.: Autodesk Inventor, 3D Studio, ADI, Advanced Modeling Extension, AME, AutoCAD, AutoCAD Mechanical Desktop, AutoCAD Development System, AutoCAD LT, Autodesk, Autodesk Animator, AutoLISP, AutoShade, AutoVision, and Heidi.
The following are trademarks of Autodesk, Inc.: ACAD, Autodesk Device Interface, AutoCAD DesignCenter, AutoTrack, Heads-up Design, ObjectARX and Visual LISP.
Microsoft, Windows are either registered trademarks or trademarks of Microsoft Corporation.
All other trademarks are trademarks of their respective holders.

Shih, Randy H.
     Parametric modeling with Autodesk Inventor R10/
Randy H. Shih

ISBN 1-58503-237-9 (Text only)
ISBN 1-58503-238-7 (Text bundled with Autodesk Inventor R10)

The author and publisher of this book have used their best efforts in preparing this book. These efforts include the development, research and testing of the material presented. The author and publisher shall not be liable in any event for incidental or consequential damages with, or arising out of, the furnishing, performance, or use of the material.

Printed and bound in the United States of America.

Preface

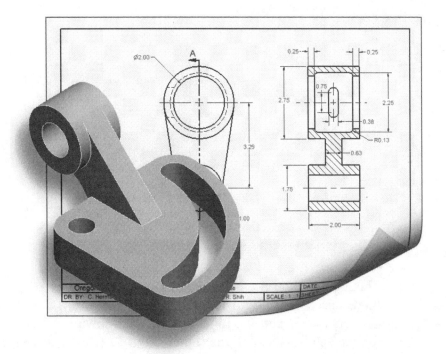

The primary goal of ***Parametric Modeling with Autodesk Inventor R10*** is to introduce the aspects of **Solid Modeling** and **Parametric Modeling**. This text is intended to be used as a training guide for students and professionals. This text covers *Autodesk Inventor R10* and the lessons proceed in a pedagogical fashion to guide you from constructing basic shapes to building intelligent solid models, creating multi-view drawings and assembly models. This text takes a hands-on, exercise-intensive approach to all the important *Parametric Modeling* techniques and concepts. This textbook contains a series of ten tutorial style lessons designed to introduce beginning CAD users to **Autodesk Inventor**. This text is also helpful to *Autodesk Inventor* users upgrading from a previous release of the software. The solid modeling techniques and concepts discussed in this text are also applicable to other parametric feature-based CAD packages. The basic premise of this book is that the more designs you create using *Autodesk Inventor*, the better you learn the software. With this in mind, each lesson introduces a new set of commands and concepts, building on previous lessons. This book does not attempt to cover all of the *Autodesk Inventor*'s features, only to provide an introduction to the software. It is intended to help you establish a good basis for exploring and growing in the exciting field of **Computer Aided Engineering**.

Acknowledgments

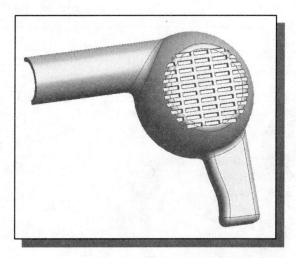

This book would not have been possible without a great deal of support. First, special thanks to two great teachers, Prof. George R. Schade of University of Nebraska-Lincoln and Mr. Denwu Lee, who showed me the fundamentals, the intrigue, and the sheer fun of Computer Aided Engineering.

The effort and support of the editorial and production staff of Schroff Development Corporation is gratefully acknowledged. I would especially like to thank Stephen Schroff and Mary Schmidt for their support and helpful suggestions during this project.

I am grateful that the Mechanical Engineering Technology Department of Oregon Institute of Technology has provided me with an excellent environment in which to pursue my interests in teaching and research.

Finally, truly unbounded thanks are due to my wife Hsiu-Ling and our daughter Casandra for their understanding and encouragement throughout this project.

Randy H. Shih
Klamath Falls, Oregon
Spring, 2005

Table of Contents

Chapter 3
Constructive Solid Geometry Concepts

Chapter 4
Model History Tree

Chapter 5
Parametric Constraints Fundamentals

Chapter 6
Geometric Construction Tools

Chapter 7
Parent/Child Relationships and the BORN Technique

Chapter 8
Part Drawings and Associative Functionality

Chapter 9
Datum Features and Auxiliary Views

Chapter 10
Symmetrical Features in Designs

Chapter 11
Advanced 3D Construction Tools

Chapter 12
Assembly Modeling - Putting It All Together

Index

NOTES:

Chapter 1
Introduction – Getting Started

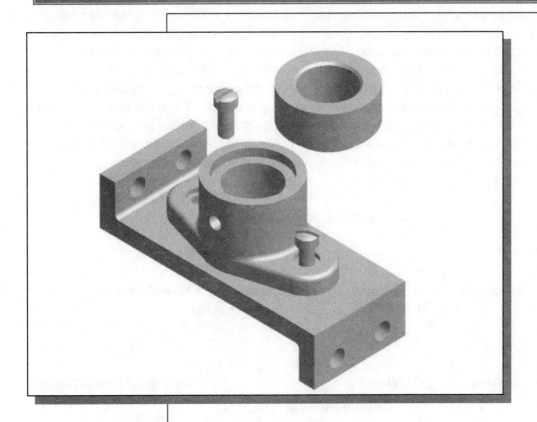

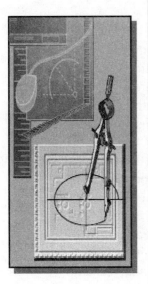

Learning Objectives

- ♦ **Development of Computer Geometric Modeling**
- ♦ **Feature-Based Parametric Modeling**
- ♦ **Startup Options and Units Setup**
- ♦ **Autodesk Inventor Screen Layout**
- ♦ **User Interface & Mouse Buttons**
- ♦ **Autodesk Inventor On-Line Help**

Introduction

The rapid changes in the field of **Computer Aided Engineering** (CAE) have brought exciting advances in the engineering community. Recent advances have made the long-sought goal of **concurrent engineering** closer to a reality. CAE has become the core of concurrent engineering and is aimed at reducing design time, producing prototypes faster, and achieving higher product quality. *Autodesk Inventor* is an integrated package of mechanical computer aided engineering software tools developed by *Autodesk, Inc. Autodesk Inventor* is a tool that facilitates a concurrent engineering approach to the design and stress-analysis of mechanical engineering products. The computer models can also be used by manufacturing equipment such as machining centers, lathes, mills, or rapid prototyping machines to manufacture the product. In this text, we will be dealing only with the solid modeling modules used for part design and part drawings.

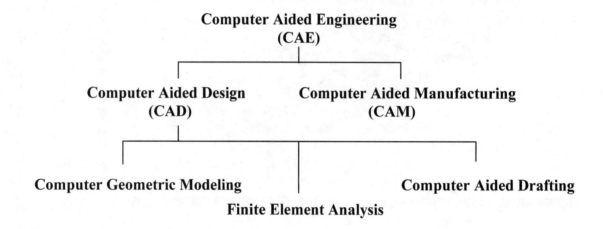

Development of Computer Geometric Modeling

Computer geometric modeling is a relatively new technology, and its rapid expansion in the last fifty years is truly amazing. Computer-modeling technology has advanced along with the development of computer hardware. The first generation CAD programs, developed in the 1950s, were mostly non-interactive; CAD users were required to create program-codes to generate the desired two-dimensional (2D) geometric shapes. Initially, the development of CAD technology occurred mostly in academic research facilities. The Massachusetts Institute of Technology, Carnegie-Mellon University, and Cambridge University were the leading pioneers at that time. The interest in CAD technology spread quickly and several major industry companies, such as General Motors, Lockheed, McDonnell, IBM, and Ford Motor Co., participated in the development of interactive CAD programs in the 1960s. Usage of CAD systems was primarily in the automotive industry, aerospace industry, and government agencies that developed their own programs for their specific needs. The 1960s also marked the beginning of the development of finite element analysis methods for computer stress analysis and computer aided manufacturing for generating machine toolpaths.

The 1970s are generally viewed as the years of the most significant progress in the development of computer hardware, namely the invention and development of **microprocessors**. With the improvement in computing power, new types of 3D CAD programs that were user-friendly and interactive became reality. CAD technology quickly expanded from very simple **computer aided drafting** to very complex **computer aided design**. The use of 2D and 3D wireframe modelers was accepted as the leading edge technology that could increase productivity in industry. The developments of surface modeling and solid modeling technologies were taking shape by the late 1970s, but the high cost of computer hardware and programming slowed the development of such technology. During this period, the available CAD systems all required room-sized mainframe computers that were extremely expensive.

In the 1980s, improvements in computer hardware brought the power of mainframes to the desktop at less cost and with more accessibility to the general public. By the mid-1980s, CAD technology had become the main focus of a variety of manufacturing industries and was very competitive with traditional design/drafting methods. It was during this period of time that 3D solid modeling technology had major advancements, which boosted the usage of CAE technology in industry.

The introduction of the *feature-based parametric solid modeling* approach, at the end of the 1980s, elevated CAD/CAM/CAE technology to a new level. In the 1990s, CAD programs evolved into powerful design/manufacturing/management tools. CAD technology has come a long way, and during these years of development, modeling schemes progressed from two-dimensional (2D) wireframe to three-dimensional (3D) wireframe, to surface modeling, to solid modeling and, finally, to feature-based parametric solid modeling.

The first generation CAD packages were simply 2D **computer aided drafting** programs, basically the electronic equivalents of the drafting board. For typical models, the use of this type of program would require that several to many views of the objects be created individually as they would be on the drafting board. The 3D designs remained in the designer's mind, not in the computer database. Mental translations of 3D objects to 2D views are required throughout the use of these packages. Although such systems have some advantages over traditional board drafting, they are still tedious and labor intensive. The need for the development of 3D modelers came quite naturally, given the limitations of the 2D drafting packages.

The development of three-dimensional modeling schemes started with three-dimensional (3D) wireframes. Wireframe models are models consisting of points and edges, which are straight lines connecting between appropriate points. The edges of wireframe models are used, similar to lines in 2D drawings, to represent transitions of surfaces and features. The use of lines and points is also a very economical way to represent 3D designs.

The development of the 3D wireframe modeler was a major leap in the area of computer geometric modeling. The computer database in the 3D wireframe modeler contains the locations of all the points in space coordinates, and it is typically sufficient to create just one model rather than multiple views of the same model. This single 3D model can then be viewed from any direction as needed. Most 3D wireframe modelers allow the user to create projected lines/edges of 3D wireframe models. In comparison to other types of 3D modelers, the 3D wireframe modelers require very little computing power and generally can be used to achieve reasonably good representations of 3D models. However, because surface definition is not part of a wireframe model, all wireframe images have the inherent problem of ambiguity. Two examples of such ambiguity are illustrated.

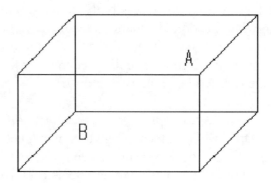

Wireframe Ambiguity: Which corner is in front, A or B?

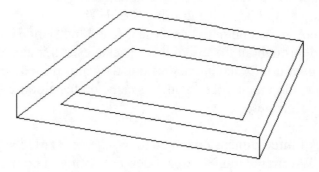

A non-realizable object: Wireframe models contain no surface definitions.

Surface modeling is the logical development in computer geometry modeling to follow the 3D wireframe modeling scheme by organizing and grouping edges that define polygonal surfaces. Surface modeling describes the part's surfaces but not its interiors. Designers are still required to interactively examine surface models to insure that the various surfaces on a model are contiguous throughout. Many of the concepts used in 3D wireframe and surface modelers are incorporated in the solid modeling scheme, but it is solid modeling that offers the most advantages as a design tool.

In the solid modeling presentation scheme, the solid definitions include nodes, edges, and surfaces, and it is a complete and unambiguous mathematical representation of a precisely enclosed and filled volume. Unlike the surface modeling method, solid modelers start with a solid or use topology rules to guarantee that all of the surfaces are stitched together properly. Two predominant methods for representing solid models are **constructive solid geometry** (CSG) representation and **boundary representation** (B-rep).

The CSG representation method can be defined as the combination of 3D solid primitives. What constitutes a "primitive" varies somewhat with the software but typically includes a rectangular prism, a cylinder, a cone, a wedge, and a sphere. Most solid modelers also allow the user to define additional primitives, which are shapes typically formed by the basic shapes. The underlying concept of the CSG representation method is very straightforward; we simply **add** or **subtract** one primitive from another. The CSG approach is also known as the machinist's approach, as it can be used to simulate the manufacturing procedures for creating the 3D object.

In the B-rep representation method, objects are represented in terms of their spatial boundaries. This method defines the points, edges, and surfaces of a volume, and/or issues commands that sweep or rotate a defined face into a third dimension to form a solid. The object is then made up of the unions of these surfaces that completely and precisely enclose a volume.

By the 1980s, a new paradigm called *concurrent engineering* had emerged. With concurrent engineering, designers, design engineers, analysts, manufacturing engineers, and management engineers all work together closely right from the initial stages of the design. In this way, all aspects of the design can be evaluated and any potential problems can be identified right from the start and throughout the design process. Using the principles of concurrent engineering, a new type of computer modeling technique appeared. The technique is known as the *feature-based parametric modeling technique*. The key advantage of the *feature-based parametric modeling technique* is its capability to produce very flexible designs. Changes can be made easily and design alternatives can be evaluated with minimum effort. Various software packages offer different approaches to feature-based parametric modeling, yet the end result is a flexible design defined by its design variables and parametric features.

Feature-Based Parametric Modeling

One of the key elements in the *Autodesk Inventor* solid modeling software is its use of the **feature-based parametric modeling technique**. The feature-based parametric modeling approach has elevated solid modeling technology to the level of a very powerful design tool. Parametric modeling automates the design and revision procedures by the use of parametric features. Parametric features control the model geometry by the use of design variables. The word *parametric* means that the geometric definitions of the design, such as dimensions, can be varied at any time during the design process. Features are predefined parts or construction tools for which users define the key parameters. A part is described as a sequence of engineering features, which can be modified/changed at any time. The concept of parametric features makes modeling more closely match the actual design-manufacturing process than the mathematics of a solid modeling program. In parametric modeling, models and drawings are updated automatically when the design is refined.

Parametric modeling offers many benefits:

- **We begin with simple, conceptual models with minimal detail; this approach conforms to the design philosophy of "shape before size."**

- **Geometric constraints, dimensional constraints, and relational parametric equations can be used to capture design intent.**

- **The ability to update an entire system, including parts, assemblies and drawings after changing one parameter of complex designs.**

- **We can quickly explore and evaluate different design variations and alternatives to determine the best design.**

- **Existing design data can be reused to create new designs.**

- **Quick design turn-around.**

One of the key features of *Autodesk Inventor* is the use of an assembly-centric paradigm, which enables users to concentrate on the design without depending on the associated parameters or constraints. Users can specify how parts fit together and the *Autodesk Inventor assembly-based fit function* automatically determines the parts' sizes and positions. This unique approach is known as the **Direct Adaptive Assembly approach**, which defines part relationships directly with no order dependency.

The *Adaptive Assembly approach* is a unique design methodology that can only be found in *Autodesk Inventor*. The goal of this methodology is to improve the design process and allows you, the designer, to **Design the Way You Think**.

Getting Started with *Autodesk Inventor*

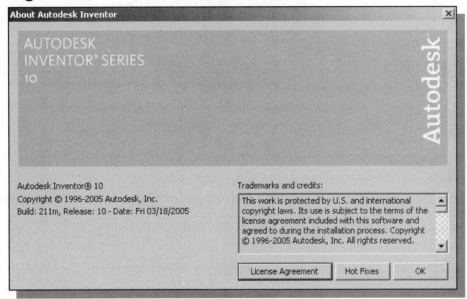

- *Autodesk Inventor* is composed of several application software modules (these modules are called *applications*), all sharing a common database. In this text, the main concentration is placed on the solid modeling modules used for part design. The general procedures required in creating solid models, engineering drawings, and assemblies are illustrated.

How to start *Autodesk Inventor* depends on the type of workstation and the particular software configuration you are using. With most *Windows* systems, you may select **Autodesk Inventor** on the *Start* menu or select the **Autodesk Inventor** icon on the desktop. Consult your instructor or technical support personnel if you have difficulty starting the software. The program takes a while to load, so be patient.

The tutorials in this text are based on the assumption that you are using *Autodesk Inventor's* default settings. If your system has been customized for other uses, contact your technical support personnel to restore the default software configuration.

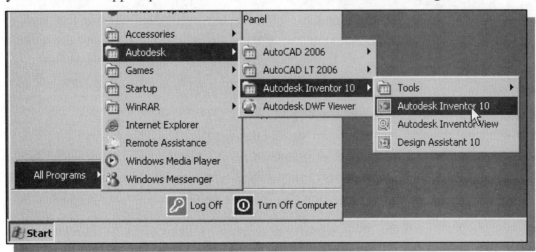

The Startup dialog box and Units Setup

Once the program is loaded into the memory, the *Startup* dialog box appears at the center of the screen.

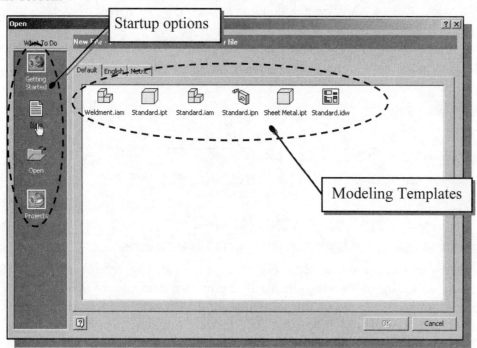

The following three startup options are available: **New, Open, Getting Started,** and **Projects.** The **New** option allows us to start a new modeling task. The **Open** option allows us to open an existing model file. The **Getting Started** option provides some quick helps that illustrate the features and general procedures of using *Autodesk Inventor*.

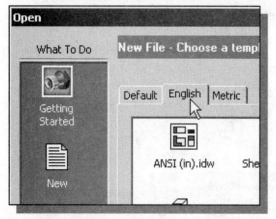

➤ Select the **New** icon with a single click of the left-mouse-button in the *What to Do* dialog box.

➤ Select the **English** tab as shown below. When starting a new CAD file, the first thing we should do is to choose the units we would like to use. We will use the English (feet and inches) setting for this example.

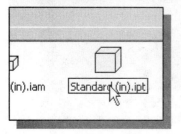

➤ Select the **Standard(in).ipt** icon as shown. The different icons are templates for the different modeling tasks. The **idw** file type stands for *Inventor* drawing file, the **iam** file type stands for *Inventor* assembly file, and the **ipt** file type stands for *Inventor* part file. The **ipn** file type stands for assembly presentation. Click **OK** in the *Startup* dialog box to accept the selected settings.

Autodesk Inventor Screen Layout

The default *Autodesk Inventor* drawing screen contains the *pull-down* menus, the *Standard* toolbar, the *Features* toolbar, the *Sketch* toolbar, the *drawing* area, the *browser* area, and the *Status Bar*. A line of quick text appears next to the icon as you move the *mouse cursor* over different icons. You may resize the *Autodesk Inventor* drawing window by click and drag at the edges of the window, or relocate the window by click and drag at the window title area.

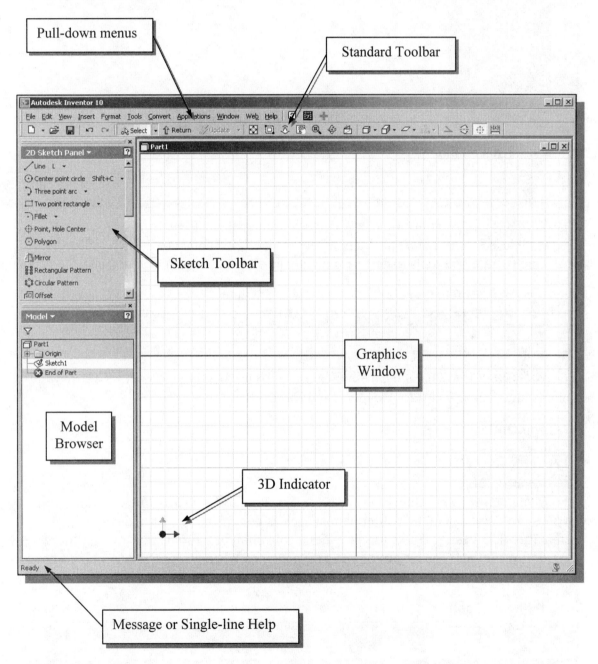

- **Pull-down Menus**

The *pull-down* menus at the top of the main window contain operations that you can use for all modes of the system.

- **Standard Toolbar**

The *Standard* toolbar at the top of the screen window allows us quick access to frequently used commands. For example, the view-related commands, such as Zoom, Pan, and Rotate, are tools to help manipulate the viewing of graphical objects.

- **Part Features Toolbar**

The *Part Features* toolbar allows us quick access to frequently used features-related commands, such as Extrude, Revolve, and Patterns.

- **Help Options**

The *Help options* icons allow us to access on-line help for *Autodesk Inventor*. The **Help topics** provides general help information, such as command options and command references. The **Visual Syllabus** provides a collection of tutorials illustrating different *Inventor* operations. The **Recover** option is activated when errors and/or problems have occurred; the system will provide useful tips on resolving the problems.

- **2D Sketch Toolbar**

The *2D Sketch* toolbar provides tools for creating the basic geometry that can be used to create features and parts.

- **Graphics Window**

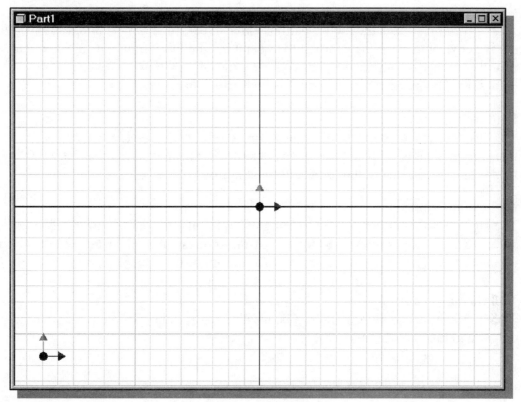

The *graphics window* is the area where models and drawings are displayed.

- **Graphics Cursor or Crosshairs**

The *graphics cursor*, or *crosshairs*, shows the location of the pointing device in the graphics window. During geometric construction, the coordinate of the cursor is displayed in the *Status Bar* area, located at the bottom of the screen. The cursor's appearance depends on the selected command or option.

- **Message and Status Bar**

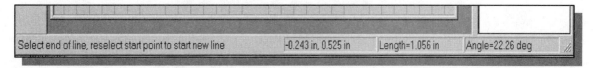

The *Message and Status Bar* area shows a single-line help when the cursor is on top of an icon. This area also displays information pertinent to the active operation. For example, in the figure above, the coordinates and length information of a line are displayed while the *Line* command is activated.

- **Browser**

The *browser window* shows and hides selected features, filters contents, manages access to features and editing, and provides alternate access to functions in the context menu.

Mouse Buttons

Autodesk Inventor utilizes the mouse buttons extensively. In learning *Autodesk Inventor*'s interactive environment, it is important to understand the basic functions of the mouse buttons. It is highly recommended that you use a mouse or a tablet with *Autodesk Inventor* since the package uses the buttons for various functions.

- **Left mouse button**
  The **left-mouse-button** is used for most operations, such as selecting menus and icons, or picking graphic entities. One click of the button is used to select icons, menus and form entries, and to pick graphic items.

- **Right mouse button**
  The **right-mouse-button** is used to bring up additional available options. The software also utilizes the **right-mouse-button** as the same as the **ENTER** key, and is often used to accept the default setting to a prompt or to end a process.

- **Middle mouse button/wheel**
  The middle mouse button/wheel can be used to Pan (hold down the wheel button and drag the mouse) or Zoom (rotate the wheel) realtime.

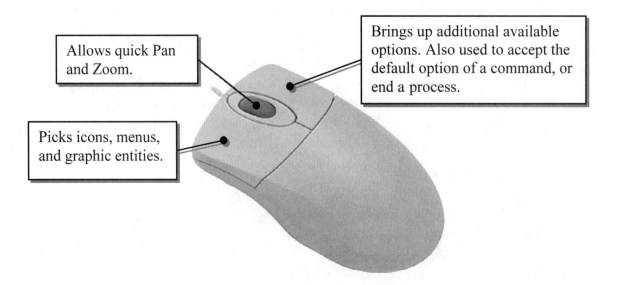

Brings up additional available options. Also used to accept the default option of a command, or end a process.

Allows quick Pan and Zoom.

Picks icons, menus, and graphic entities.

[Esc] - Canceling Commands

The [**Esc**] key is used to cancel a command in *Autodesk Inventor*. The [**Esc**] key is located near the top-left corner of the keyboard. Sometimes, it may be necessary to press the [**Esc**] key twice to cancel a command; it depends on where we are in the command sequence. For some commands, the [**Esc**] key is used to exit the command.

On-Line Help

❖ Several types of on-line help are available at any time during an *Autodesk Inventor* session. *Autodesk Inventor* provides many on-line help functions, such as:

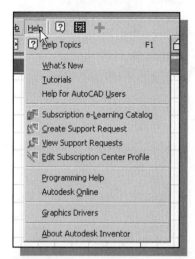

- The **Help** menu: Click on the **Help** option in the pull-down menu to access the *Autodesk Inventor* **Help** menu **system**.

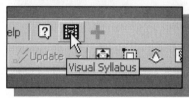

- On-line manuals and tutorials: Two icons are available in the *Standard* toolbar to access the on-line manuals and tutorials: **Help Topics** and **Visual Syllabus**.

- Help quick key: Press the [**F1**] key to access the *On-line Help* system.

Leaving Autodesk Inventor

➢ To leave *Autodesk Inventor*, use the left-mouse-button and click on **File** at the top of the *Autodesk Inventor* screen window, then choose **Exit** from the pull-down menu.

Creating a CAD Files Folder

It is a good practice to create a separate folder to store your CAD files. You should not save your CAD files in the same folder where the *Autodesk Inventor* application is located. It is much easier to organize and backup your project files if they are in a separate folder. Making folders within this folder for different types of projects will help you organize your CAD files even further. When creating CAD files in *Autodesk Inventor*, it is strongly recommended that you *save* your CAD files on the hard drive.

➢ To create a new folder in the *Windows* environment:

1. In *My Computer*, or start *Windows Explorer* under the *Start* menu, open the folder in which you want to create a new folder.

2. On the **File** menu, point to **New**, and then click **Folder**. The new folder appears with a temporary name.

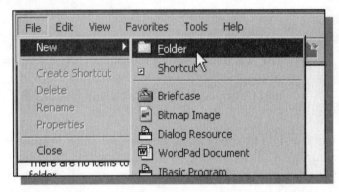

3. Type a name for the new folder, and then press **ENTER**.

Chapter 2
Parametric Modeling Fundamentals

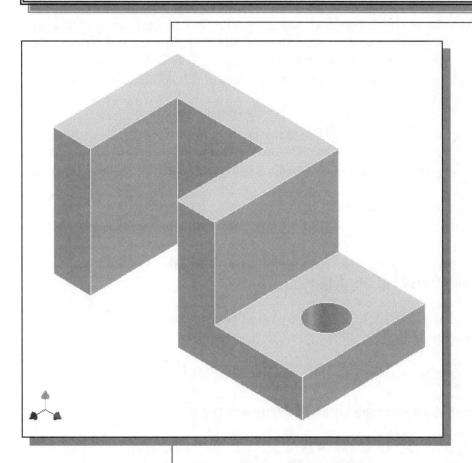

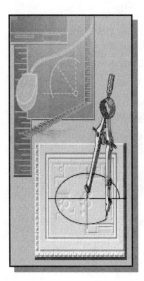

Learning Objectives

- ♦ **Create Simple Extruded Solid Models**
- ♦ **Understand the Basic Parametric Modeling Procedure**
- ♦ **Create 2-D Sketches**
- ♦ **Understand the "Shape before Size" Approach**
- ♦ **Use the Dynamic Viewing Commands**
- ♦ **Create and Edit Parametric Dimensions**

Introduction

The **feature-based parametric modeling** technique enables the designer to incorporate the original **design intent** into the construction of the model. The word ***parametric*** means the geometric definitions of the design, such as dimensions, can be varied at any time in the design process. Parametric modeling is accomplished by identifying and creating the key features of the design with the aid of computer software. The design variables, described in the sketches and described as parametric relations, can then be used to quickly modify/update the design.

In *Autodesk Inventor*, the parametric part modeling process involves the following steps:

1. **Create a rough two-dimensional sketch of the basic shape of the base feature of the design.**

2. **Apply/modify constraints and dimensions to the two-dimensional sketch.**

3. **Extrude, revolve, or sweep the parametric two-dimensional sketch to create the base solid feature of the design.**

4. **Add additional parametric features by identifying feature relations and complete the design.**

5. **Perform analyses on the computer model and refine the design as needed.**

6. **Create the desired drawing views to document the design.**

The approach of creating two-dimensional sketches of the three-dimensional features is an effective way to construct solid models. Many designs are in fact the same shape in one direction. Computer input and output devices we use today are largely two-dimensional in nature, which makes this modeling technique quite practical. This method also conforms to the design process that helps the designer with conceptual design along with the capability to capture the ***design intent***. Most engineers and designers can relate to the experience of making rough sketches on restaurant napkins to convey conceptual design ideas. *Autodesk Inventor* provides many powerful modeling and design-tools, and there are many different approaches to accomplishing modeling tasks. The basic principle of **feature-based modeling** is to build models by adding simple features one at a time. In this chapter, the general parametric part modeling procedure is illustrated; a very simple solid model with extruded features is used to introduce the *Autodesk Inventor* user interface. The display viewing functions and the basic two-dimensional sketching tools are also demonstrated.

The *Adjuster* design

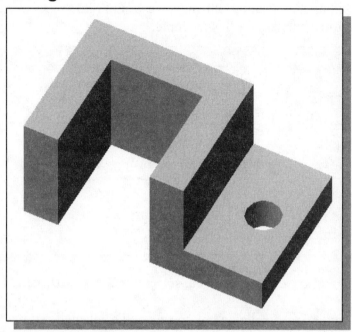

Starting *Autodesk Inventor*

1. Select the **Autodesk Inventor** option on the *Start* menu or select the **Autodesk Inventor** icon on the desktop to start *Autodesk Inventor*. The *Autodesk Inventor* main window will appear on the screen.

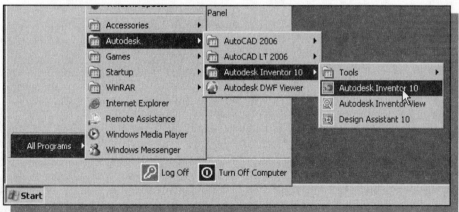

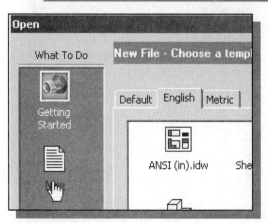

2. Select the **New** icon with a single click of the left-mouse-button in the *What to Do* dialog box.

3. Select the **English** tab as shown below. When starting a new CAD file, the first thing we should do is choose the units we would like to use. We will use the English setting (inches) for this example.

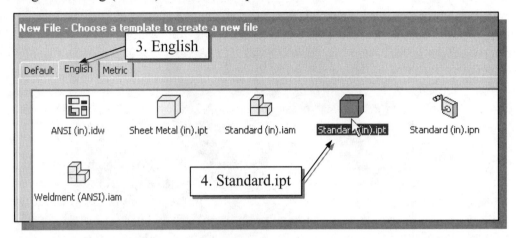

4. In the *New - Choose Template* area, select the **Standard(in).ipt** icon as shown.

5. Pick **OK** in the *Startup* dialog box to accept the selected settings.

Autodesk Inventor Screen Layout

➢ The default *Autodesk Inventor* drawing screen contains the *pull-down* menus, the *Standard* toolbar, the *Sketch* toolbar, the *graphics* window, the *browser* area, and the *Status Bar*.

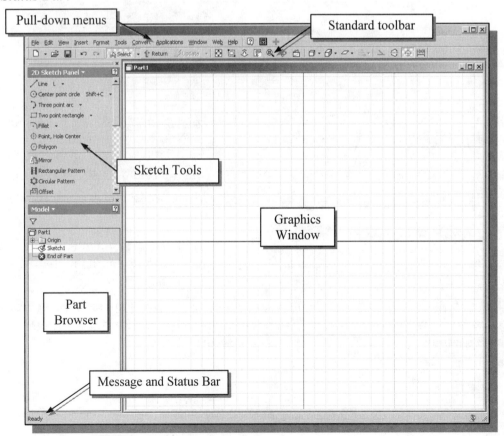

Creating Rough Sketches

Quite often during the early design stage, the shape of a design may not have any precise dimensions. Most conventional CAD systems require the user to input the precise lengths and locations of all geometric entities defining the design, which are not available during the early design stage. With *parametric modeling*, we can use the computer to elaborate and formulate the design idea further during the initial design stage. With *Autodesk Inventor*, we can use the computer as an electronic sketchpad to help us concentrate on the formulation of forms and shapes for the design. This approach is the main advantage of *parametric modeling* over conventional solid-modeling techniques.

As the name implies, a **rough sketch** is not precise at all. When sketching, we simply sketch the geometry so that it closely resembles the desired shape. Precise scale or lengths are not needed. *Autodesk Inventor* provides us with many tools to assist us in finalizing sketches. For example, geometric entities such as horizontal and vertical lines are set automatically. However, if the rough sketches are poor, it will require much more work to generate the desired parametric sketches. Here are some general guidelines for creating sketches in *Autodesk Inventor*:

- **Create a sketch that is proportional to the desired shape.** Concentrate on the shapes and forms of the design.

- **Keep the sketches simple.** Leave out small geometry features such as fillets, rounds and chamfers. They can easily be placed using the Fillet and Chamfer commands after the parametric sketches have been established.

- **Exaggerate the geometric features of the desired shape.** For example, if the desired angle is 85 degrees, create an angle that is 50 or 60 degrees. Otherwise, *Autodesk Inventor* might assume the intended angle to be a 90-degree angle.

- **Draw the geometry so that it does not overlap.** The geometry should eventually form a closed region. *Self-intersecting* geometry shapes are not allowed.

- **The sketched geometric entities should form a closed region.** To create a solid feature, such as an extruded solid, a closed region is required so that the extruded solid forms a 3D volume.

- ➤ **Note:** The concepts and principles involved in *parametric modeling* are very different, and sometimes they are totally opposite, to those of conventional computer aided drafting. In order to understand and fully utilize *Autodesk Inventor's* functionality, it will be helpful to take a *Zen* approach to learning the topics presented in this text: **Temporarily forget your knowledge and experiences of using conventional Computer Aided Drafting systems.**

Step 1: Creating a Rough Sketch

➢ The *Sketch* toolbar provides tools for creating the basic geometry that can be used to create features and parts.

1. Move the graphics cursor to the **Line** icon in the *Sketch* toolbar. A *Help-tip box* appears next to the cursor and a brief description of the command is displayed at the bottom of the drawing screen: *"Creates Straight line segments and tangent arcs."*

2. Select the icon by clicking once with the **left-mouse-button**; this will activate the Line command. In the command prompt area, near the bottom of the *Autodesk Inventor* drawing screen, the message *"Specify start point, drag off endpoint for tangent arcs"* is displayed. *Autodesk Inventor* expects us to identify the starting location of a straight line.

Graphics Cursors

Notice the cursor changes from an arrow to a crosshair when graphical input is expected.

1. Left-click a starting point for the shape, roughly near the lower center of the graphics window as shown below.

2. As you move the graphics cursor, you will see a digital readout in the *Status Bar* area at the bottom of the window. The readout gives you the cursor location, the line length, and the angle of the line measured from horizontal. Move the cursor around and you will notice different symbols appear at different locations.

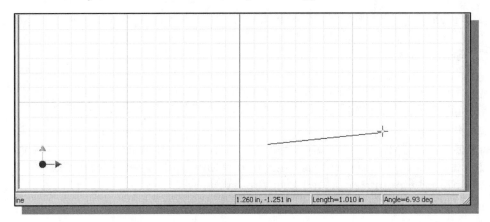

3. Move the graphics cursor toward the right side of the graphics window and create a horizontal line as shown below (**Point 2**). Notice the geometric constraint symbol displayed.

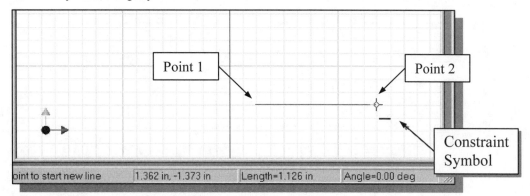

Geometric Constraint Symbols

Autodesk Inventor displays different visual clues, or symbols, to show you alignments, perpendicularities, tangencies, etc. These constraints are used to capture the *design intent* by creating constraints where they are recognized. *Autodesk Inventor* displays the governing geometric rules as models are built. To prevent constraints from forming, hold down the [**Ctrl**] key while creating an individual sketch curve. For example, while sketching line segments with the Line command, endpoints are joined with a *coincident constraint*, but when the [**Ctrl**] key is pressed and held, the inferred constraint will not be created.

‖	**Vertical**	indicates a line is vertical
≡	**Horizontal**	indicates a line is horizontal
– – –	**Dashed line**	indicates the alignment is to the center point or endpoint of an entity
⫽	**Parallel**	indicates a line is parallel to other entities
⊾	**Perpendicular**	indicates a line is perpendicular to other entities
⌒	**Coincident**	indicates the cursor is at the endpoint of an entity
⊙	**Concentric**	indicates the cursor is at the center of an entity
⟳	**Tangent**	indicates the cursor is at tangency points to curves

4. Complete the sketch as shown below, creating a closed region ending at the starting point (Point 1.) Do not be overly concerned with the actual size of the sketch. Note that all line segments are sketched horizontally or vertically.

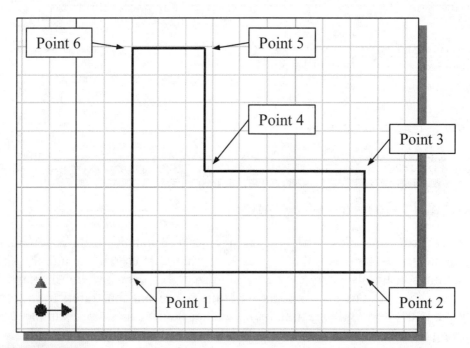

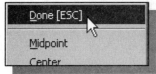

5. Inside the graphics window, click once with the **right-mouse-button** to display the option menu. Select Done[Esc] in the popup menu, or hit the [**Esc**] key once, to end the Sketch Line command.

Step 2: Apply/Modify Constraints and Dimensions

➢ As the sketch is made, *Autodesk Inventor* automatically applies some of the geometric constraints (such as horizontal, parallel, and perpendicular) to the sketched geometry. We can continue to modify the geometry, apply additional constraints, and/or define the size of the existing geometry. In this example, we will illustrate adding dimensions to describe the sketched entities.

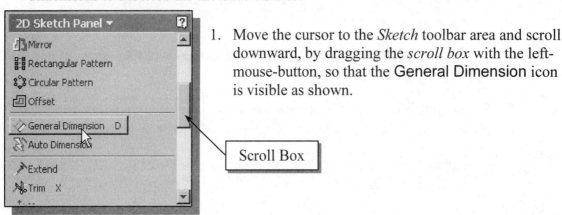

1. Move the cursor to the *Sketch* toolbar area and scroll downward, by dragging the *scroll box* with the left-mouse-button, so that the General Dimension icon is visible as shown.

2. Move the cursor on top of the **General Dimension** icon. The **General Dimension** command allows us to quickly create and modify dimensions. Left-click once on the icon to activate the **General Dimension** command.

3. The message "*Select Geometry to Dimension*" is displayed in the *Status Bar* area at the bottom of the *Inventor* window. Select the bottom-right horizontal line by left-clicking once on the line.

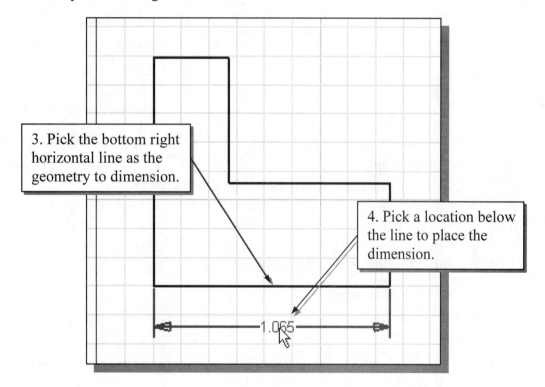

3. Pick the bottom right horizontal line as the geometry to dimension.

4. Pick a location below the line to place the dimension.

4. Move the graphics cursor below the selected line and left-click to place the dimension. (Note that the value displayed on your screen might be different than what is shown in the figure above.)

5. The message "*Select Geometry to Dimension*" is displayed in the *Status Bar* area, at the bottom of the *Inventor* window. Select the lower right-vertical line.

6. Pick a location toward the right of the sketch to place the dimension.

❖ The **General Dimension** command will create a length dimension if a single line is selected.

7. The message "*Select Geometry to Dimension*" is displayed in the *Status Bar* area, at the bottom of the *Inventor* window. Select the top-horizontal line as shown below.

8. Select the bottom-horizontal line as shown below.

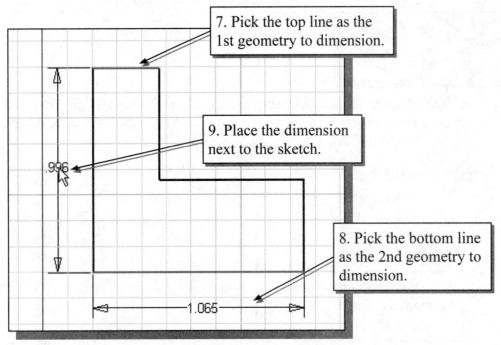

7. Pick the top line as the 1st geometry to dimension.

9. Place the dimension next to the sketch.

8. Pick the bottom line as the 2nd geometry to dimension.

9. Pick a location to the left of the sketch to place the dimension.

❖ When two parallel lines are selected, the **General Dimension** command will create a dimension measuring the distance between them.

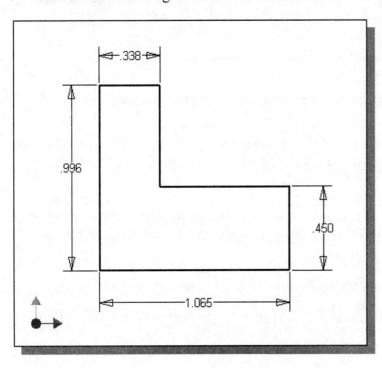

10. On you own, repeat the above steps and create additional dimensions so that the sketch appears as shown.

Modifying the Dimensions of the Sketch

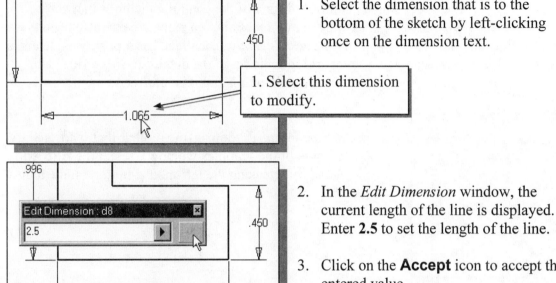

1. Select the dimension that is to the bottom of the sketch by left-clicking once on the dimension text.

2. In the *Edit Dimension* window, the current length of the line is displayed. Enter **2.5** to set the length of the line.

3. Click on the **Accept** icon to accept the entered value.

➢ *Autodesk Inventor* will now update the profile with the new dimension value.

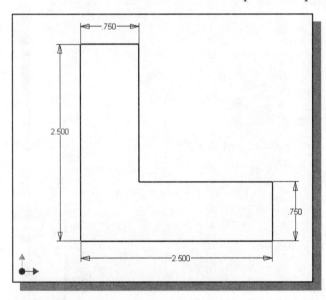

4. On you own, repeat the above steps and adjust the dimensions so that the sketch appears as shown.

5. Inside the graphics window, click once with the **right-mouse-button** to display the option menu. Select **Done** in the popup menu to end the General Dimension command.

6. Inside the graphics window, click once with the **right-mouse-button** and select **Finish Sketch** in the popup menu to end the Sketch option.

Step 3: Completing the Base Solid Feature

Now that the 2D sketch is completed, we will proceed to the next step: create a 3D part from the 2D profile. Extruding a 2D profile is one of the common methods that can be used to create 3D parts. We can extrude planar faces along a path. We can also specify a height value and a tapered angle. In *Autodesk Inventor*, each face has a positive side and a negative side, the current face we're working on is set as the default positive side. This positive side identifies the positive extrusion direction and it is referred to as the face's ***normal***.

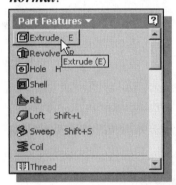

1. In the *Part Features* toolbar (the toolbar that is located to the left side of the graphics window), select the **Extrude** command by releasing the left-mouse-button on the icon.

2. In the *Extrude* popup window, enter **2.5** as the extrusion distance. Notice that the sketch region is automatically selected as the extrusion profile.

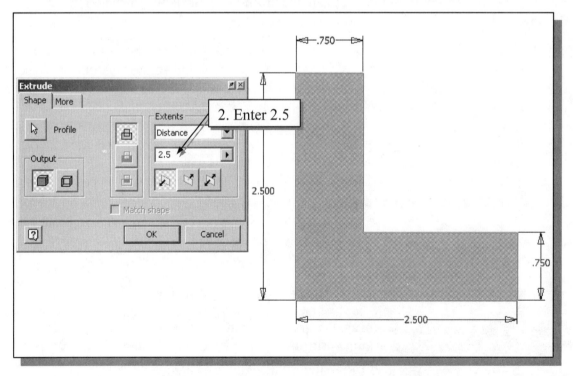

3. Click on the **OK** button to proceed with creating the 3D part.

➤   Note that all dimensions disappeared from the screen. All parametric definitions are stored in the ***Autodesk Inventor* database** and any of the parametric definitions can be displayed and edited at any time.

Isometric View

❖ *Autodesk Inventor* provides many ways to display views of the three-dimensional design. Several options are available that allow us to quickly view the design to track the overall effect of any changes being made to the model. We will first orient the model to display in the *isometric view*, by using the pull-down menu.

1. Select **Isometric View** in the **View** pull-down menu to change the display to the isometric view. (Note that **F6** can also be used to activate this command.)

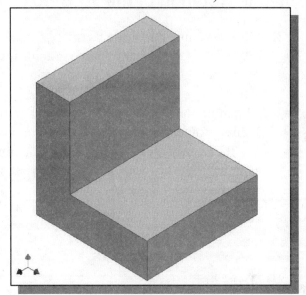

❖ Notice the other view-related commands that are available under the **View** pull-down menu. Most of these commands are also accessible through the *Standard* toolbar and/or *function keys*.

Dynamic Viewing Functions – *Zoom* and *Pan*

• *Autodesk Inventor* provides a special user interface called *Dynamic Viewing* that enables convenient viewing of the entities in the graphics window.

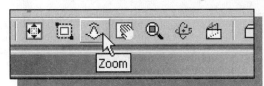

1. Click on the **Zoom** icon, located in the *Standard* toolbar as shown.

2. Move the cursor near the center of the graphics window.

3. Inside the graphics window, **press and hold down the left-mouse-button**, then move downward to enlarge the current display scale factor.

4. Press the [**Esc**] key once to exit the Zoom command.

5. Click on the **Pan** icon, located next to the Zoom command in the *Standard* toolbar. The icon is the picture of a hand.

> ➤ The Pan command enables us to move the view to a different position. This function acts as if you are using a video camera.

6. On your own, use the Zoom and Pan options to reposition the sketch near the center of the screen.

Dynamic Rotation of the 3D Block – *3D Rotate*

The 3D Rotate command allows us to:
*   Rotate a part or assembly in the graphics window. Rotation can be around the center mark, free in all directions, or around the X/Y-axes in the *3D-Rotate* display.
*   Reposition the part or assembly in the graphics window.
*   Display isometric or standard orthographic views of a part or assembly.
*   The 3D Rotate tool is accessible while other tools are active. *Autodesk Inventor* remembers the last used mode when you exit the Rotate command.

1. Click on the **Rotate** icon in the *Standard* toolbar.

The *3D Rotate* display is a circular rim with four handles and a center mark. *3D Rotate* enables us to manipulate the view of 3D objects by clicking and dragging with the left-mouse-button:

*   Drag with the left-mouse-button near the center for free rotation.
*   Drag on the handles to rotate around the horizontal or vertical axes.
*   Drag on the rim to rotate about an axis that is perpendicular to the displayed view.
*   Single left-mouse-click to align the center mark of the view.

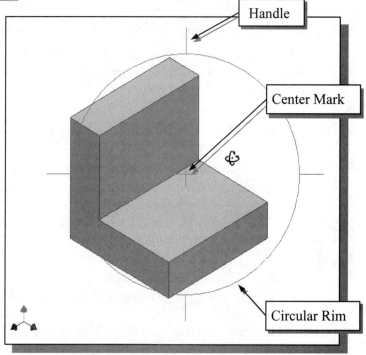

2. Inside the *circular rim*, press down the left-mouse-button and drag in an arbitrary direction; the 3D Rotate command allows us to freely rotate the solid model.

3. Move the cursor near the circular rim and notice the cursor symbol changes to a single circle. Drag with the left-mouse-button to rotate about an axis that is perpendicular to the displayed view.

4. Single left-mouse-click near the top-handle to align the selected location to the center mark in the graphics window.

5. Press the **[Space bar]** to activate the **Common Views** option.

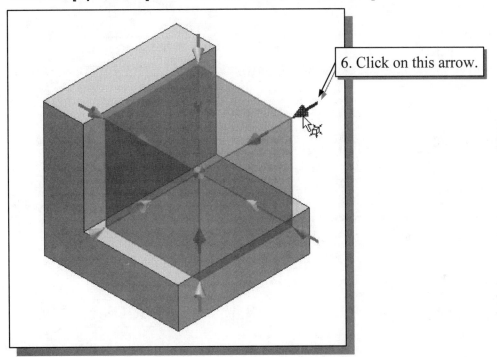

6. Click on this arrow.

6. The displayed arrows represent the viewing directions of the pre-defined common views. Change the viewing direction by left-clicking once on the top-right arrow as shown in the figure above. Hit the **[Esc]** key once to exit the 3D Rotate command.

7. On your own, use the different options described in the above steps and familiarize yourself with the 3D Rotate command. Reset the display to the *Isometric* view as shown in the above figure before continuing to the next section.

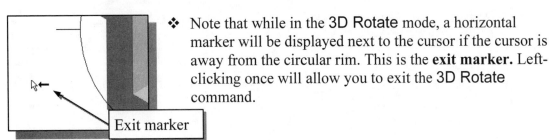

Exit marker

❖ Note that while in the 3D Rotate mode, a horizontal marker will be displayed next to the cursor if the cursor is away from the circular rim. This is the **exit marker.** Left-clicking once will allow you to exit the 3D Rotate command.

Dynamic Viewing – Quick Keys

We can also use the function keys on the keyboard and the mouse to access the *Dynamic Viewing* functions.

❖ **Panning – F2 and the mouse**

> Hold the **F2** function key down, and drag with the left-mouse-button to pan the display. This allows you to reposition the display while maintaining the same scale factor of the display.

Pan   F2   +   MOUSE

❖ **Zooming – F3 and the mouse**

> Hold the **F3** function key down, and drag with the left-mouse-button vertically on the screen to adjust the scale of the display. Moving upward will reduce the scale of the display, making the entities display smaller on the screen. Moving downward will magnify the scale of the display.

Zoom   F3   +   MOUSE

❖ **3D Dynamic Rotation – F4 and the mouse**

> Hold the **F4** function key down and move the mouse to rotate the display. The 3D Rotate rim with four handles and the center mark appears on the screen. Note that the Common View option is not available when using the **F4** quick key.

Dynamic Rotation   F4   +   MOUSE

Display Modes

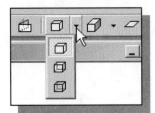

- The display in the graphics window has three basic display-modes: wireframe, shaded, and shaded with hidden edge display. To change the display mode in the active window, click on the triangle icon next to the display mode button on the *Standard* toolbar, as shown in the figure.

❖ Shaded Solid:

The first icon in the display mode button group generates a shaded image of the 3D object.

❖ Hidden-Edge Display:

The second icon in the display mode button group can be used to generate an image of the 3D object with all the back lines shown.

❖ Wireframe Image:

The third icon in the display mode button group allows the display of the 3D objects using the basic wireframe representation scheme.

Orthographic vs. Perspective

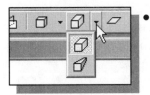

- Besides the three basic display modes, we can also choose orthographic view or perspective view of the display. Click on the icon next to the display mode button on the *Standard* toolbar, as shown in the figure.

❖ Orthographic Camera

The first icon allows the display of the 3D object using the orthographics, parallel edges, representation scheme.

❖ Perspective Camera

The second icon allows the display of the 3D object using the perspective, nonparallel edges representation scheme.

➢ On your own, use the different options described in the above sections to familiarize yourself with the 3D viewing/display commands. Reset the display to the standard **isometric view** before continuing to the next section.

Sketch Plane – It is an XY CRT, but an XYZ World

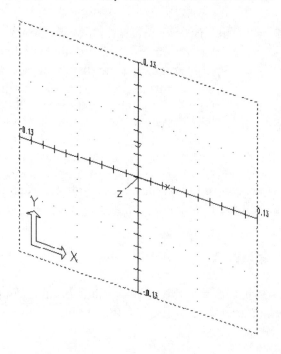

Design modeling software is becoming more powerful and user friendly, yet the system still does only what the user tells it to do. When using a geometric modeler, we therefore need to have a good understanding of what its inherent limitations are. We should also have a good understanding of what we want to do and what to expect, as the results are based on what is available.

In most 3D geometric modelers, 3D objects are located and defined in what is usually called **world space** or **global space**. Although a number of different coordinate systems can be used to create and manipulate objects in a 3D modeling system, the objects are typically defined and stored using the world space. The world space is usually a **3D Cartesian coordinate system** that the user cannot change or manipulate.

In most engineering designs, models can be very complex, and it would be tedious and confusing if only the world coordinate system were available. Practical 3D modeling systems allow the user to define **Local Coordinate Systems (LCS)** or **User Coordinate Systems (UCS)** relative to the world coordinate system. Once a local coordinate system is defined, we can then create geometry in terms of this more convenient system.

Although objects are created and stored in 3D space coordinates, most of the geometry entities can be referenced using 2D Cartesian coordinate systems. Typical input devices such as a mouse or digitizers are two-dimensional by nature; the movement of the input device is interpreted by the system in a planar sense. The same limitation is true of common output devices, such as CRT displays and plotters. The modeling software performs a series of three-dimensional to two-dimensional transformations to correctly project 3D objects onto a 2D picture plane.

The *Autodesk Inventor* **sketch plane** is a special construction tool that enables the planar nature of 2D input devices to be directly mapped into the 3D coordinate system. The *sketch plane* is a local coordinate system that can be aligned to the world coordinate system, an existing face of a part, or a reference plane. By default, the *sketch plane* is aligned to the world coordinate system.

Think of a sketch plane as the surface on which we can sketch the 2D profiles of the parts. It is similar to a piece of paper, a white board, or a chalkboard that can be attached to any planar surface. The first profile we create is usually drawn on the default sketch plane, which is in the current coordinate system. Subsequent profiles can then be drawn on sketch planes that are defined on **planar faces of a part**, **work planes attached to part geometry**, or **sketch planes attached to a coordinate system** (such as the World XY, XZ, and YZ sketch planes). The model we have created so far used the default settings where the sketch plane is aligned to the XY plane of the world coordinate system.

1. In the *Standard* toolbar select the **Sketch** command by left-clicking once on the icon.

2. In the *Status Bar* area, the message: "*Select face, workplane, sketch or sketch geometry*" is displayed. *Autodesk Inventor* expects us to identify a planar surface where the 2D sketch of the next feature is to be created. Move the graphics cursor on the 3D part and notice that *Autodesk Inventor* will automatically highlight feasible planes and surfaces as the cursor is on top of the different surfaces. Pick the top horizontal face of the 3D solid object.

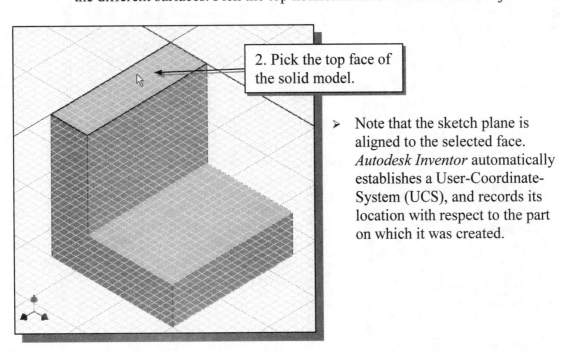

2. Pick the top face of the solid model.

➢ Note that the sketch plane is aligned to the selected face. *Autodesk Inventor* automatically establishes a User-Coordinate-System (UCS), and records its location with respect to the part on which it was created.

Step 4-1: Adding an Extruded Feature

- Next, we will create and profile another sketch, a rectangle, which will be used to create another extrusion feature that will be added to the existing solid object.

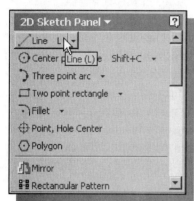

1. Select the **Line** command by clicking once with the **left-mouse-button** on the icon in the *Sketch* toolbar.

2. Create a sketch with segments perpendicular/parallel to the existing edges of the solid model as shown below.

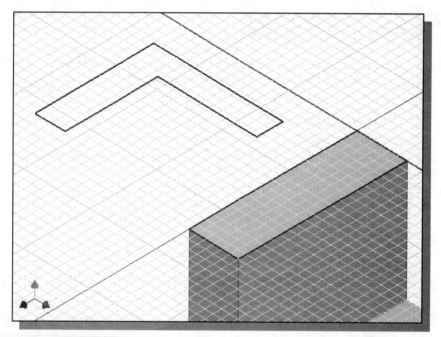

3. Select the **General Dimension** command in the *Sketch* toolbar. The General Dimension command allows us to quickly create and modify dimensions. Left-click once on the icon to activate the General Dimension command.

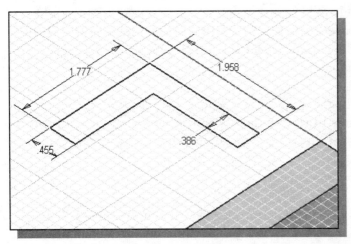

4. The message *"Select Geometry to Dimension"* is displayed in the *Status Bar* area, at the bottom of the *Inventor* window. Create the four dimensions to describe the size of the sketch as shown in the figure.

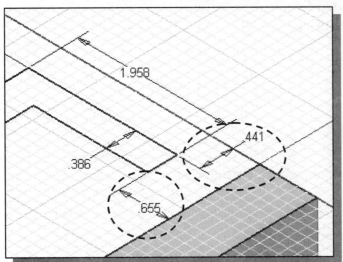

5. Create the two location dimensions to describe the position of the sketch relative to the top corner of the solid model as shown.

6. On your own, modify the two location dimensions to **0.0** and the size dimensions as shown in the below figure.

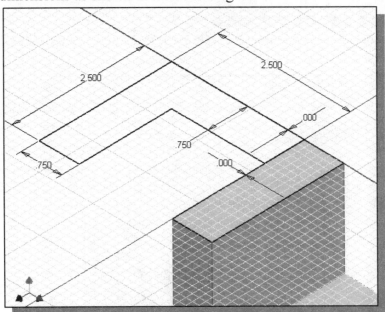

7. Inside the graphics window, click once with the **right-mouse-button** to display the option menu. Select **Done** in the popup menu to end the General Dimension command.

8. Inside the graphics window, click once with the **right-mouse-button** to display the option menu. Select **Finish Sketch** in the popup menu to end the Sketch option.

9. In the *Part Features* toolbar (the toolbar that is located to the left side of the graphics window), select the **Extrude** command by releasing the left-mouse-button on the icon.

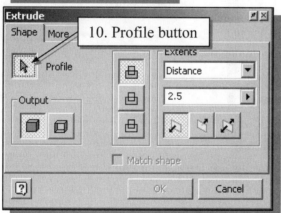

10. In the *Extrude* popup window, the **Profile** button is pressed down; *Autodesk Inventor* expects us to identify the profile to be extruded.

11. Move the cursor inside the rectangle we just created and left-click once to select the region as the profile to be extruded.

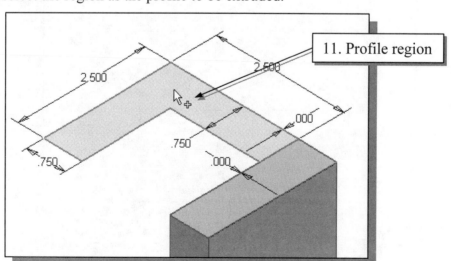

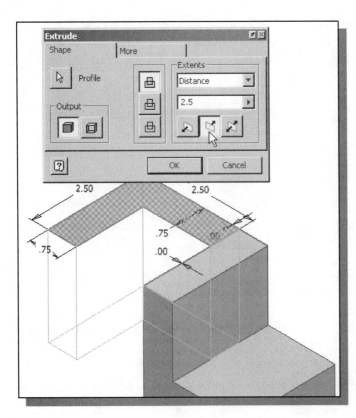

12. In the *Extrude* popup window, enter **2.5** as the extrude distance as shown.

13. Click on the second **direction icon** to set the extrusion direction downward as shown.

14. Click on the **OK** button to proceed with creating the extruded feature.

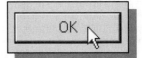

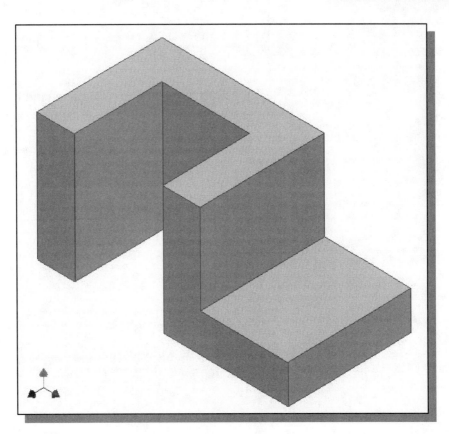

Step 4-2: Adding a Cut Feature

- Next, we will create and profile a circle, which will be used to create a **cut** feature that will be added to the existing solid object.

1. In the *Standard* toolbar select the **Sketch** command by left-clicking once on the icon.

2. In the *Status Bar* area, the message: "*Select face, workplane, sketch or sketch geometry.*" is displayed. *Autodesk Inventor* expects us to identify a planar surface where the 2D sketch of the next feature is to be created. Pick the top horizontal face of the 3D solid model as shown.

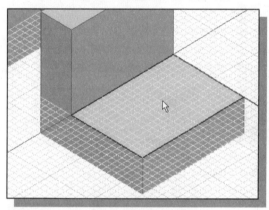

> Note that the sketch plane is aligned to the selected face. *Autodesk Inventor* automatically establishes a User-Coordinate-System (UCS), and records its location with respect to the part on which it was created.

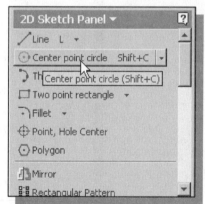

3. Select the **Center point circle** command by clicking once with the **left-mouse-button** on the icon in the *Sketch* toolbar.

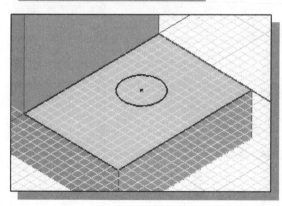

4. Create a circle of arbitrary size on the top face of the solid model as shown.

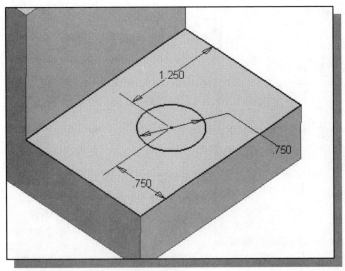

5. On your own, create and modify the dimensions of the sketch as shown in the figure.

6. Inside the graphics window, click once with the **right-mouse-button** to display the option menu. Select **Done** in the popup menu to end the General Dimension command.

7. Inside the graphics window, click once with the **right-mouse-button** to display the option menu. Select **Finish Sketch** in the popup menu to end the Sketch option.

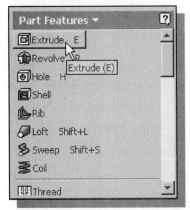

8. In the *Part Features* toolbar (the toolbar that is located to the left side of the graphics window), select the **Extrude** command by releasing the left-mouse-button on the icon.

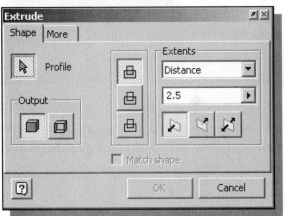

9. In the *Extrude* popup window, the **Profile** button is pressed down; *Autodesk Inventor* expects us to identify the profile to be extruded.

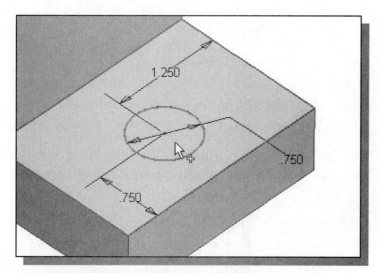

10. Click on the inside of the sketched circle as shown.

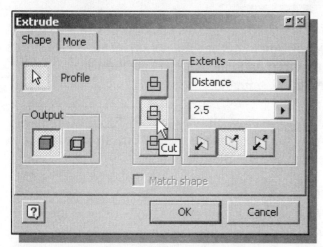

11. Click on the **CUT** icon, as shown, to set the extrusion operation to **cut**.

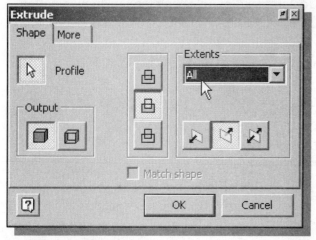

12. Set the *Extents* option to **All** as shown.

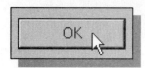

13. Click on the **OK** button to proceed with creating the extruded feature.

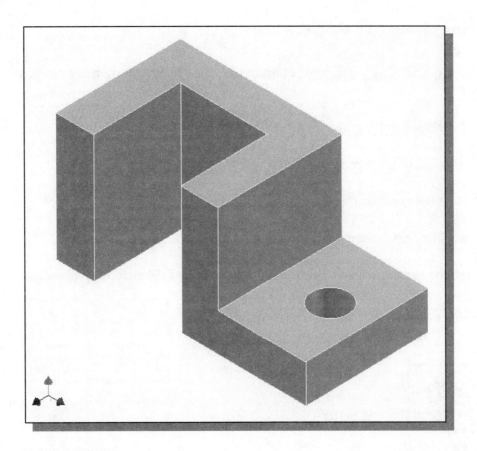

Save the Model

1. Select **Save** in the *Standard* toolbar, or you can also use the "**Ctrl-S**" combination (hold down the "**Ctrl**" key and hit the "S" key once) to save the part.

2. In the popup window, select the directory to store the model in and enter **Adjuster** as the name of the file.

3. Click on the **Save** button to save the file.

❖ You should form a habit of saving your work periodically, just in case something might go wrong while you are working on it. In general, one should save one's work at an interval of every 15 to 20 minutes. One should also save before making any major modifications to the model.

Questions:

1. What is the first thing we should set up in *Autodesk Inventor* when creating a new model?

2. Describe the general *parametric modeling* procedure.

3. What is the main difference between a rough sketch and a *profile*?

4. List two of the geometric constraint symbols used by *Autodesk Inventor*.

5. What was the first feature we created in this lesson?

6. Describe the steps required to define the orientation of the sketching plane?

7. Identify the following commands:

(a)

(b)

(c)

(d)

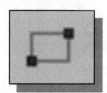

Exercises: (All dimensions are in inches.)

1.   Plate Thickness: **.25**

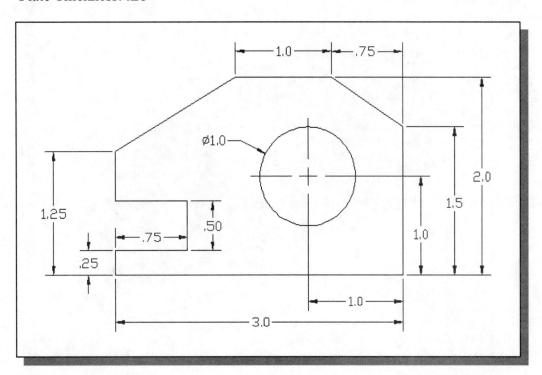

2.   Plate Thickness: **.5**

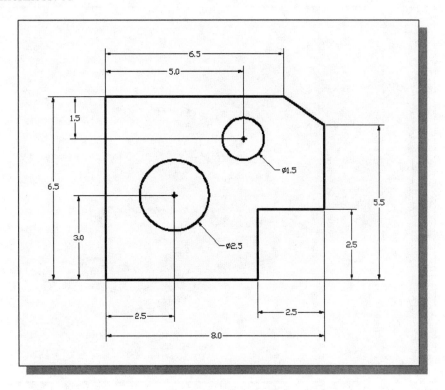

3.

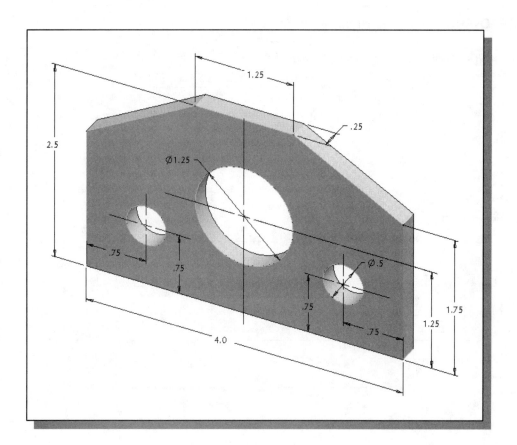

4.

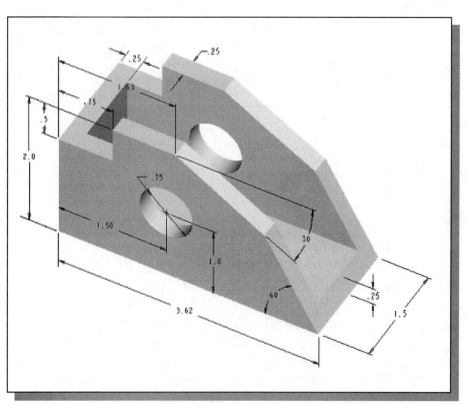

Chapter 3
Constructive Solid Geometry Concepts

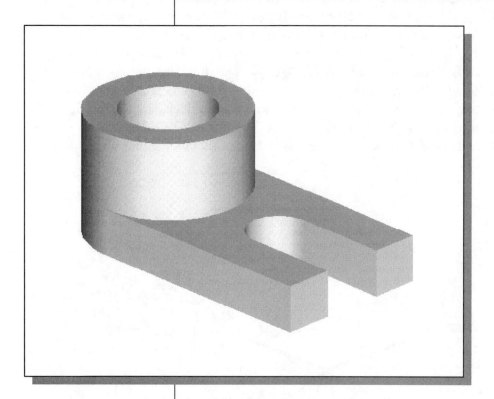

Learning Objectives

- ♦ **Understand Constructive Solid Geometry Concepts**
- ♦ **Create a Binary Tree**
- ♦ **Understand the Basic Boolean Operations**
- ♦ **Setup GRID and SNAP Intervals**
- ♦ **Understand the Importance of Order of Features**
- ♦ **Create Placed Features**
- ♦ **Use the Different Extrusion Options**

Introduction

In the 1980s, one of the main advancements in **solid modeling** was the development of the **Constructive Solid Geometry** (CSG) method. CSG describes the solid model as combinations of basic three-dimensional shapes (**primitive solids**). The basic primitive solid set typically includes: Rectangular-prism (Block), Cylinder, Cone, Sphere, and Torus (Tube). Two solid objects can be combined into one object in various ways using operations known as **Boolean operations**. There are three basic Boolean operations: **JOIN (Union)**, **CUT (Difference)**, and **INTERSECT**. The *JOIN* operation combines the two volumes included in the different solids into a single solid. The *CUT* operation subtracts the volume of one solid object from the other solid object. The *INTERSECT* operation keeps only the volume common to both solid objects. The CSG method is also known as the **Machinist's Approach**, as the method is parallel to machine shop practices.

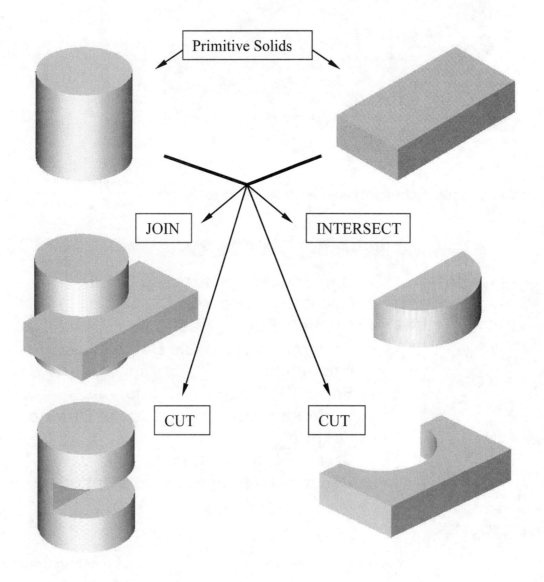

Binary Tree

The CSG is also referred to as the method used to store a solid model in the database. The resulting solid can be easily represented by what is called a **binary tree**. In a binary tree, the terminal branches (leaves) are the various primitives that are linked together to make the final solid object (the root). The binary tree is an effective way to keep track of the *history* of the resulting solid. By keeping track of the history, the solid model can be re-built by relinking through the binary tree. This provides a convenient way to modify the model. We can make modifications at the appropriate links in the binary tree and relink the rest of the history tree without building a new model.

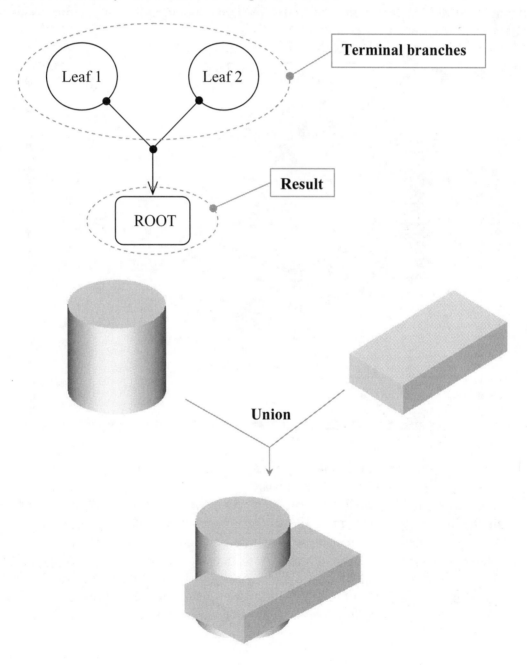

The *Locator* Design

The CSG concept is one of the important building blocks for feature-based modeling. In *Autodesk Inventor*, the CSG concept can be used as a planning tool to determine the number of features that are needed to construct the model. It is also a good practice to create features that are parallel to the manufacturing process required for the design. With parametric modeling, we are no longer limited to using only the predefined basic solid shapes. In fact, any solid features we create in *Autodesk Inventor* are used as primitive solids; parametric modeling allows us to maintain full control of the design variables that are used to describe the features. In this lesson, a more in-depth look at the parametric modeling procedure is presented. The equivalent CSG operation for each feature is also illustrated.

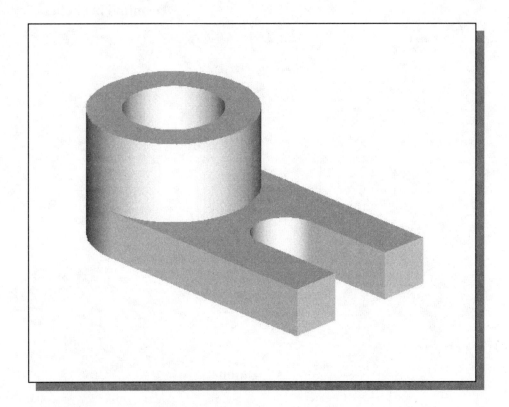

> ➤ Before going through the tutorial, on your own, make a sketch of a CSG binary tree of the **Locator** design using only two basic types of primitive solids: cylinder and rectangular prism. In your sketch, how many *Boolean operations* will be required to create the model? What is your choice of the first primitive solid to use, and why? Take a few minutes to consider these questions and do the preliminary planning by sketching on a piece of paper. Compare the sketch you make to the CSG binary tree steps shown on page 2-7. Note that there are many different possibilities in combining the basic primitive solids to form the solid model. Even for the simplest design, it is possible to take several different approaches to creating the same solid model.

Modeling Strategy – CSG Binary Tree

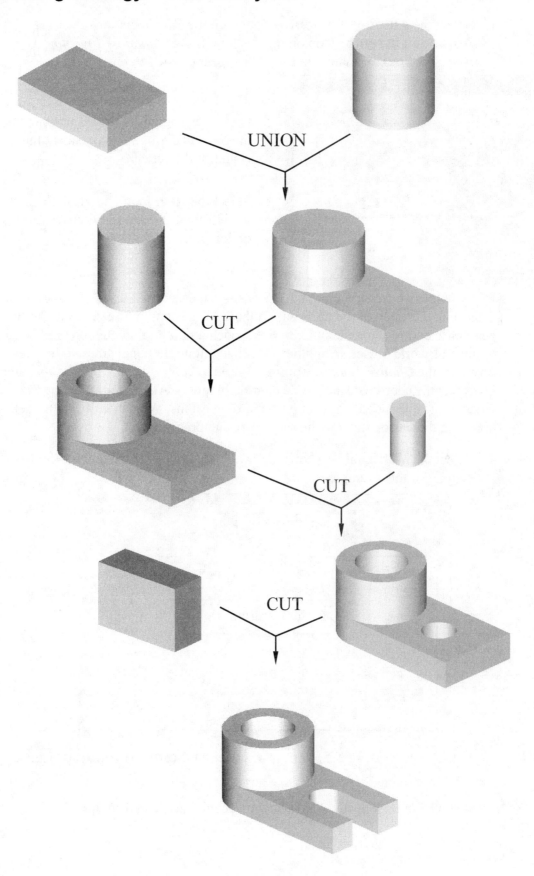

Starting Autodesk Inventor

1. Select the **Autodesk Inventor** option on the *Start* menu or select the **Autodesk Inventor** icon on the desktop to start *Autodesk Inventor*. The *Autodesk Inventor* main window will appear on the screen.

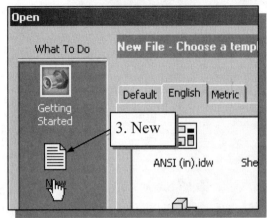

2. Once the program is loaded into memory, the *Startup* dialog box appears at the center of the screen.

3. Select the **New** icon with a single click of the left-mouse-button in the *What to Do* dialog box.

❖ Every object we construct in a CAD system is measured in units. We should determine the value of the units within the CAD system before creating the first geometric entities. For example, in one model, a unit might equal one millimeter of the real-world object; in another model, a unit might equal an inch. In *Autodesk Inventor*, the ***Choose Template*** option is used to control how *Autodesk Inventor* interprets the coordinate and angle entries. In most CAD systems, setting the model units does not always set units for dimensions. We generally set model units and dimension units to the same type and precision.

4. Select the **Metric** tab as shown below. We will use the millimeter (mm) setting for this example.

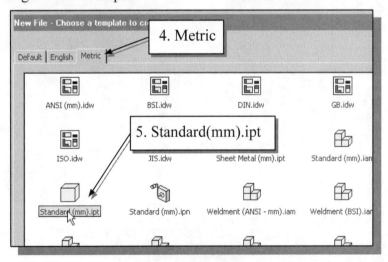

5. In the *New - Choose Template* area, select the **Standard(mm).ipt** icon as shown.

6. Pick **OK** in the *Startup* dialog box to accept the selected settings.

GRID and *SNAP* Intervals Setup

1. In the *pull-down* menu, select

   [Tools] → [Document Settings]

2. In the *Document Settings* dialog box, click on the **Sketch** tab as shown in the below figure.

3. In the *Document Settings* dialog box, set the X and Y *Snap Spacing* to **5 mm**.

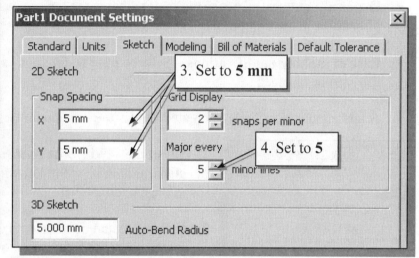

4. Change *Grid Display* to display one *major line* every **5** *minor lines*.

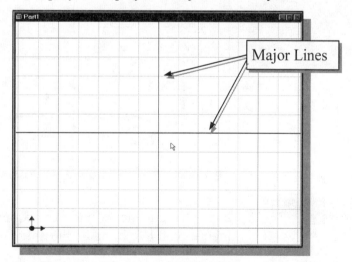

5. Pick **OK** to exit the *Sketch Settings* dialog box.

➤ Note that the above *Snap Spacing* settings actually set the grid spacing in *Autodesk Inventor*. Although the **Snap to grid** option is also available in *Autodesk Inventor*, its usage in parametric modeling is not recommended.

Base Feature

In *parametric modeling*, the first solid feature is called the **base feature,** which usually is the primary shape of the model. Depending upon the design intent, additional features are added to the base feature.

Some of the considerations involved in selecting the base feature are:

- **Design intent** – Determine the functionality of the design; identify the feature that is central to the design.

- **Order of features** – Choose the feature that is the logical base in terms of the order of features in the design.

- **Ease of making modifications** – Select a base feature that is more stable and is less likely to be changed.

➢ A rectangular block will be created first as the base feature of the ***Locator*** design.

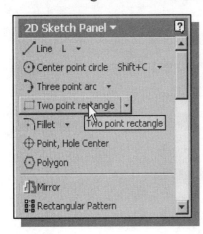

1. Select the **Two point rectangle** command by clicking once with the **left-mouse-button** on the icon in the *Sketch* toolbar.

2. Create a rectangle of arbitrary size by selecting two locations on the screen as shown below.

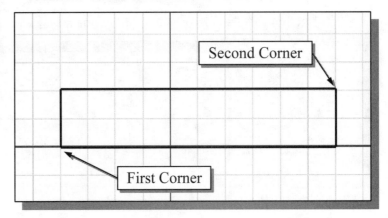

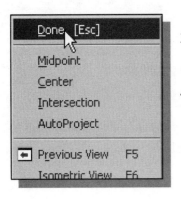

3. Inside the graphics window, click once with the right-mouse-button to bring up the option menu.

4. Select **Done** to end the Rectangle command.

5. Move the cursor to the *Sketch* toolbar area and scroll downward by dragging the *scroll box* with the left-mouse-button.

6. Move the cursor on top of the **General Dimension** icon. The General Dimension command allows us to quickly create and modify dimensions. Left-click once on the icon to activate the General Dimension command.

7. The message "*Select Geometry to Dimension*" is displayed in the *Status Bar* area at the bottom of the *Inventor* window. Select the bottom horizontal line by left-clicking once on the line.

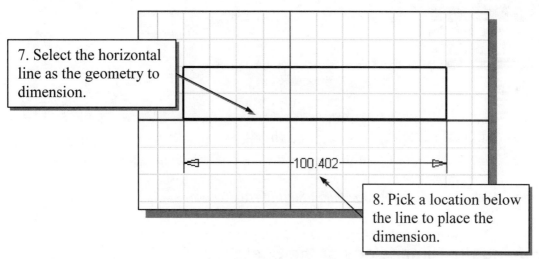

7. Select the horizontal line as the geometry to dimension.

8. Pick a location below the line to place the dimension.

8. Move the graphics cursor below the selected line and left-click to place the dimension. (Note that the value displayed on your screen might be different than what is shown in the above figure.)

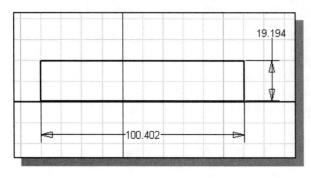

9. On your own, create the vertical size dimension of the sketched rectangle as shown.

Model Dimensions Format

1. In the pull-down menu, select

   [Tools] → [Document Settings]

2. In the *Document Settings* dialog box, set the *Modeling Dimension Display* to **Display as value** as shown in the figure.

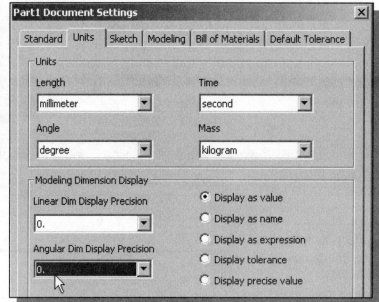

3. Also set the precision to *no digits* after the decimal point for both the linear dimension and angular dimension displays as shown in the above figure.

4. Pick **OK** to exit the *Document Settings* dialog box.

Modifying the Dimensions of the Sketch

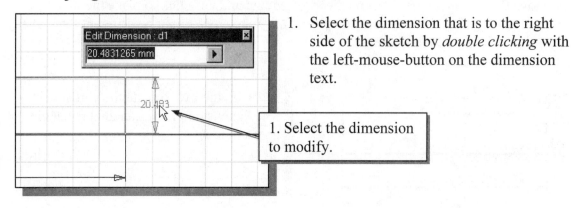

1. Select the dimension that is to the right side of the sketch by *double clicking* with the left-mouse-button on the dimension text.

1. Select the dimension to modify.

2. In the *Edit Dimension* window, the current length of the line is displayed. Enter **50** to set the selected length of the sketch to 50 millimeters.

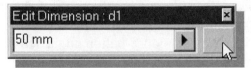

3. Click on the **Accept** icon to accept the entered value.

➢ *Autodesk Inventor* will now update the profile with the new dimension value.

4. On you own, repeat the above steps and adjust the dimensions so that the sketch appears as shown below. Also exit the **Dimension** command.

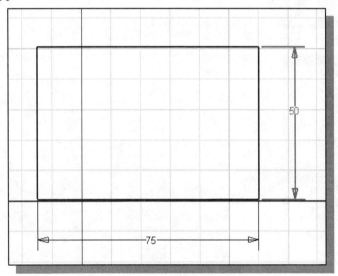

Repositioning Dimensions

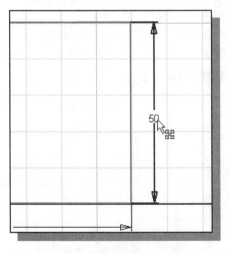

1. Move the cursor near the vertical dimension; note that the dimension is highlighted. Move the cursor slowly until a small marker appears next to the cursor, as shown in the figure.

2. Drag with the left-mouse-button to reposition the selected dimension.

3. Repeat the above steps to reposition the horizontal dimension.

4. Inside the graphics window, click once with the **right-mouse-button** to display the option menu. Select **Finish Sketch** in the popup menu to end the Sketch option.

Completing the Base Solid Feature

1. In the *Part Features* toolbar (the toolbar that is located to the left side of the graphics window), select the **Extrude** command by releasing the left-mouse-button on the icon.

2. In the *Extrude* popup window, enter **15** as the extrusion distance. Notice that the sketch region is automatically selected as the extrusion profile.

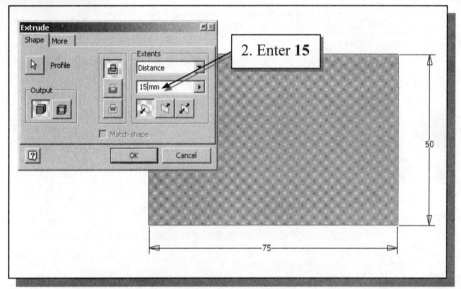

3. Click on the **OK** button to proceed with creating the 3D part. Use the *Dynamic Viewing* options to view the created part. Press F6 to change the display to the isometric view as shown before going to the next section.

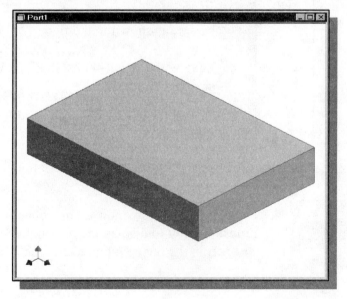

Creating the Next Solid Feature

1. In the *Standard* toolbar select the **Sketch** command by left-clicking once on the icon.

2. In the *Status Bar* area, the message: "*Select face, workplane, sketch or sketch geometry*" is displayed. *Autodesk Inventor* expects us to identify a planar surface where the 2D sketch of the next feature is to be created. Move the graphics cursor on the 3D part and notice that *Autodesk Inventor* will automatically highlight feasible planes and surfaces as the cursor is on top of the different surfaces.

3. Use the dynamic rotate function to display the bottom face of the solid model as shown below.

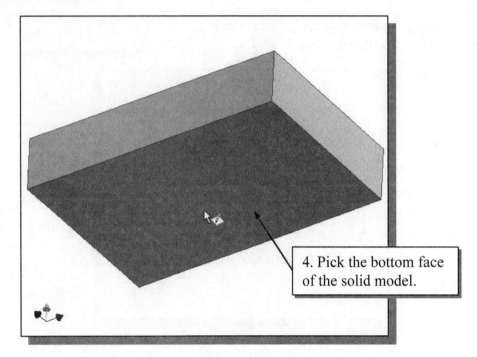

4. Pick the bottom face of the solid model.

4. Pick the bottom face of the 3D model as the sketching plane.

> Note that the sketching plane is aligned to the selected face. *Autodesk Inventor* automatically establishes a User-Coordinate-System (UCS) and records its location with respect to the part on which it was created.

5. Select the **Center point circle** command by clicking once with the left-mouse-button on the icon in the *Sketch* toolbar.

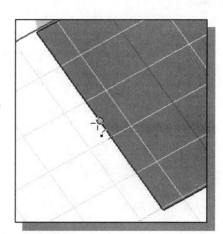

➢ We will align the center of the circle to the midpoint of the base feature.

6. Move the cursor along the shorter edge of the base feature and pick the midpoint of the edge when the midpoint is displayed with a **GREEN** color.

7. Select the corner of the base feature to create a circle as shown below.

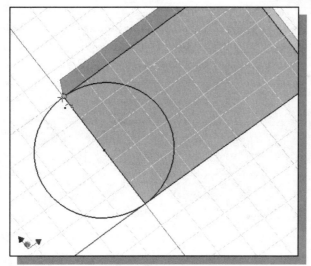

8. Inside the graphics window, click once with the right-mouse-button to display the option menu. Select **Done** in the popup menu to end the Circle command.

9. Inside the graphics window, click once with the right-mouse-button to display the option menu. Select **Finish Sketch** in the popup menu to end the Sketch option.

10. Press the function key **F6** once or select **Isometric** in the **View** pull-down menu to change the display to the isometric view as shown.

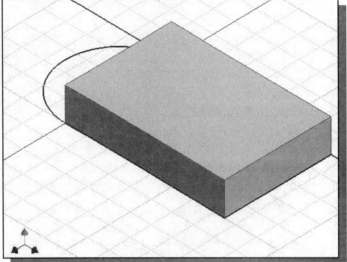

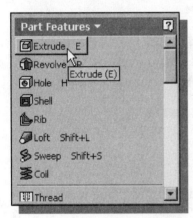

11. In the *Part Features* toolbar (the toolbar that is located to the left side of the graphics window), select the **Extrude** command by releasing the left-mouse-button on the icon.

12. In the *Extrude* popup window, the **Profile** button is pressed down; *Autodesk Inventor* expects us to identify the profile to be extruded. Move the cursor inside the circle we just created and left-click once to select the region as the profile to be extruded.

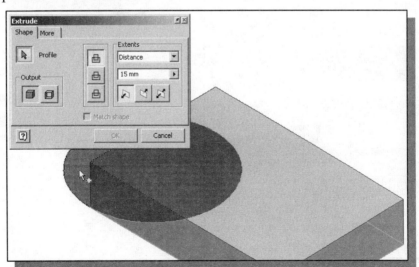

13. In the *Extrude* popup window, enter **40** as the blind extrusion distance as shown below. Confirm the operation option is set to **Join.** Click on the **Flip** button to reverse the direction of extrusion (upward) as shown below.

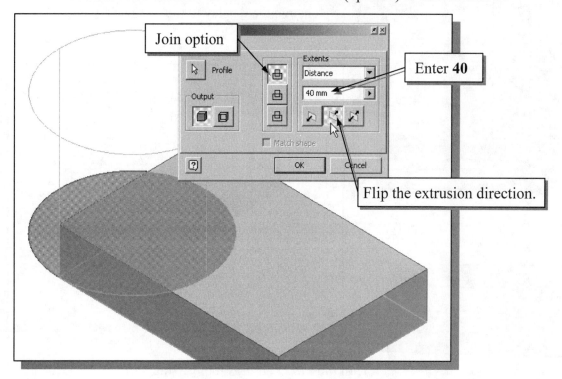

14. Click on the **OK** button to proceed with the *Join* operation.

- The two features are joined together into one solid part; the *CSG-Union* operation was performed.

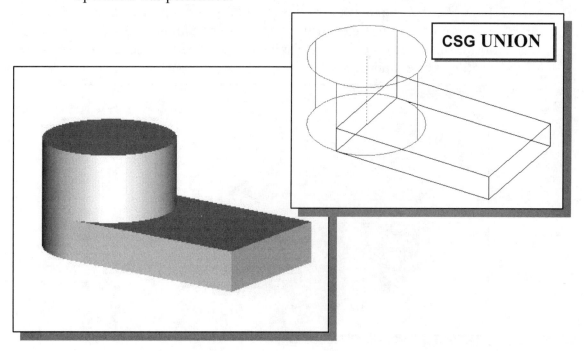

Creating a CUT Feature

• We will create a circular cut as the next solid feature of the design. We will align the sketch plane to the top of the last cylinder feature.

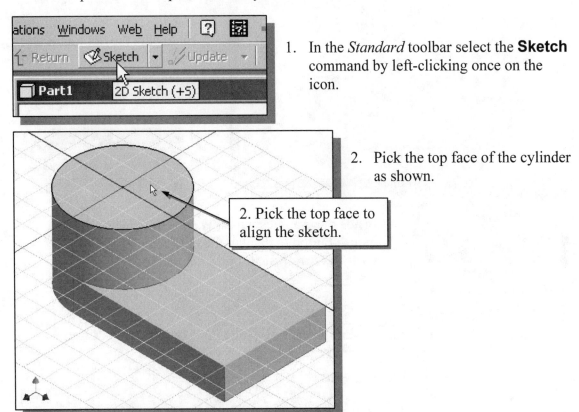

1. In the *Standard* toolbar select the **Sketch** command by left-clicking once on the icon.

2. Pick the top face of the cylinder as shown.

2. Pick the top face to align the sketch.

3. Select the **Center point circle** command by clicking once with the **left-mouse-button** on the icon in the *Sketch* toolbar.

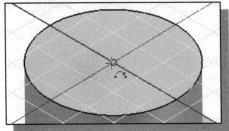

4. Select the **Center** point of the top face of the 3D model as shown.

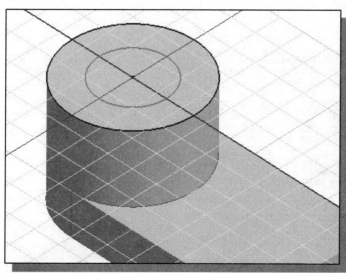

5. Sketch a circle of arbitrary size inside the top face of the cylinder as shown below.

6. Use the right-mouse-button to display the option menu and select **Done** in the popup menu to end the Circle command.

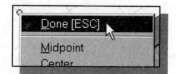

7. Inside the graphics window, click once with the right-mouse-button to display the option menu. Select the **Create Dimension** option in the popup menu.

8. Create a dimension to describe the size of the circle and set it to **30mm**.

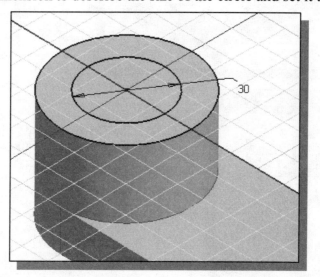

9. Inside the graphics window, click once with the right-mouse-button to display the option menu. Select **Done** in the popup menu to end the Circle command.

10. Inside the graphics window, click once with the right-mouse-button to display the option menu. Select **Finish Sketch** in the popup menu to end the Sketch option.

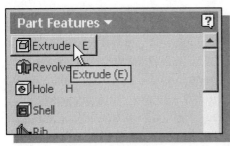

11. In the *Part Features* toolbar (the toolbar that is located to the left side of the graphics window), select the **Extrude** command by releasing the left-mouse-button on the icon.

12. In the *Extrude* popup window, the **Profile** button is pressed down; *Autodesk Inventor* expects us to identify the profile to be extruded. Move the cursor **inside** the **circle** we just created and left-click once to select the inside region as the profile to be extruded.

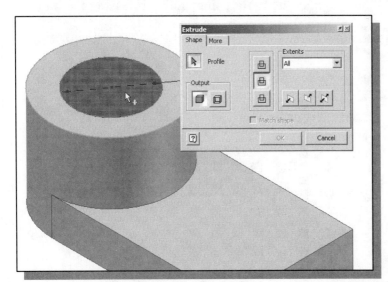

13. In the *Extrusion* popup window, set the operation option to **Cut**. Select **All** as the *Extents* option, as shown below. Confirm the arrowhead points downward.

14. Click on the **OK** button to proceed with the *Cut* operation.

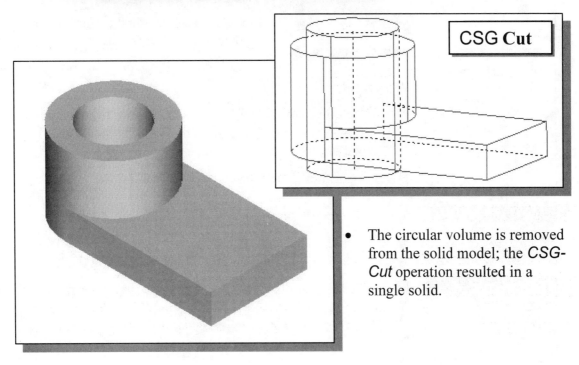

- The circular volume is removed from the solid model; the *CSG-Cut* operation resulted in a single solid.

Creating a PLACED FEATURE

- In *Autodesk Inventor*, there are two types of geometric features: **placed features** and **sketched features**. The last cut feature we created is a *sketched feature*, where we created a rough sketch and performed an extrusion operation. We can also create a hole feature, which is a placed feature. A *placed feature* is a feature that does not need a sketch and can be created automatically. Holes, fillets, chamfers, and shells are all placed features.

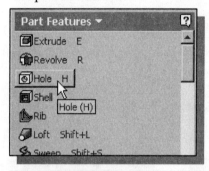

1. In the *Part Features* toolbar (the toolbar that is located to the left side of the graphics window), select the **Hole** command by releasing the left-mouse-button on the icon.

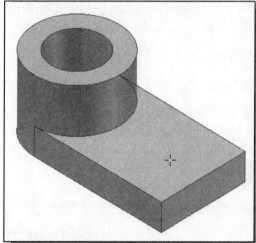

2. In the *Holes* window, notice the **Face** button is pressed down.

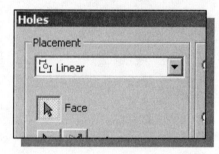

3. Pick a location inside the horizontal surface of the base feature as shown.

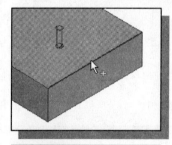

4. Pick the right-edge of the top face of the base feature as shown. This will be used as the 1st reference for placing the hole on the plane.

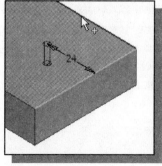

5. Pick the adjacent-edge of the top face as shown. This will be used as the 2nd reference for placing the hole on the plane.

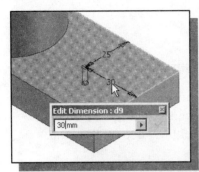

6. Click and adjust the two numbers to **25 mm** and **30 mm** as shown.

7. In *Holes* dialog box, set the *Termination* option to **Through All**.

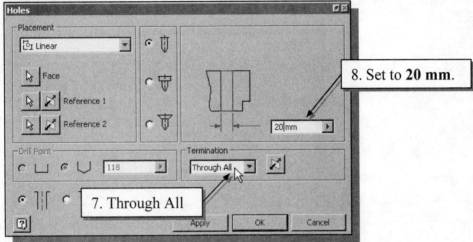

8. In the ***Drill Size*** option window, click on the diameter and change the displayed value to **20 mm** as the new diameter of the drill.

9. Click on the **OK** button to proceed with the ***Hole*** feature.

• The circular volume is removed from the solid model; the *CSG-Cut* operation resulted in a single solid.

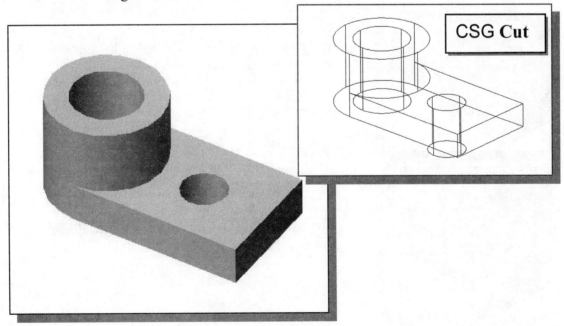

Creating a Rectangular Cut Feature

- We will create a rectangular cut as the last solid feature of the *Locator*.

1. In the *Standard* toolbar, select the **Sketch** command by left-clicking once on the icon.

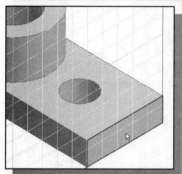

2. Pick the right face of the base feature as shown.

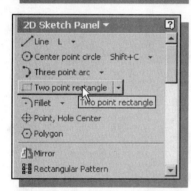

3. Select the **Two point rectangle** command by clicking once with the left-mouse-button on the icon in the *Sketch* toolbar.

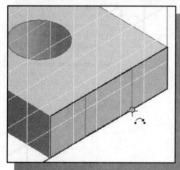

4. Create a rectangle that is aligned to the top and bottom edges of the base feature as shown.

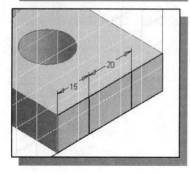

5. On your own, create and modify the two dimensions as shown.

6. Inside the graphics window, click once with the **right-mouse-button** to display the option menu. Select **Finish Sketch** in the popup menu to end the Sketch option.

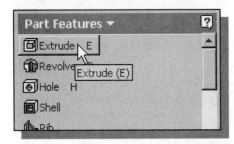

7. In the *Part Features* toolbar (the toolbar that is located to the left side of the graphics window), select the **Extrude** command by releasing the left-mouse-button on the icon.

8. In the *Extrude* popup window, the **Profile** button is pressed down; *Autodesk Inventor* expects us to identify the profile to be extruded. Move the cursor inside the rectangle we just created and left-click once to select the region as the profile to be extruded.

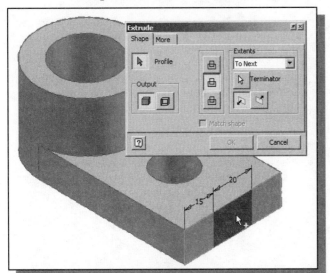

9. In the *Extrude* popup window, set the operation option to **Cut.** Select **To Next** as the *Extents* option as shown. Confirm the arrowhead points toward the center of the solid model.

10. Click on the **OK** button to create the *cut feature* and complete the design.

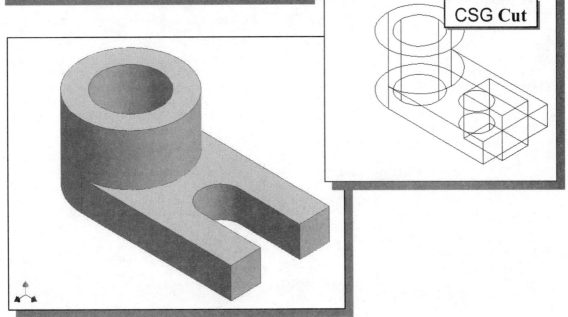

CSG Cut

Questions:

1. List and describe three basic *Boolean operations* commonly used in computer geometric modeling software?

2. What is a *primitive solid*?

3. What does *CSG* stand for?

4. Which *Boolean operation* keeps only the volume common to the two solid objects?

5. What is the main difference between a *CUT feature* and a *HOLE feature* in *Autodesk Inventor*?

6. Using the CSG concepts, create *Binary Tree* sketches showing the steps you plan to use to create the two models shown on the next page:

Ex.1)

Ex.2)

Exercises: (All dimensions are in inches.)

1.

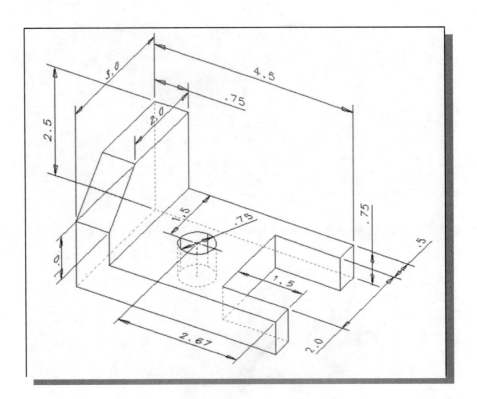

2.

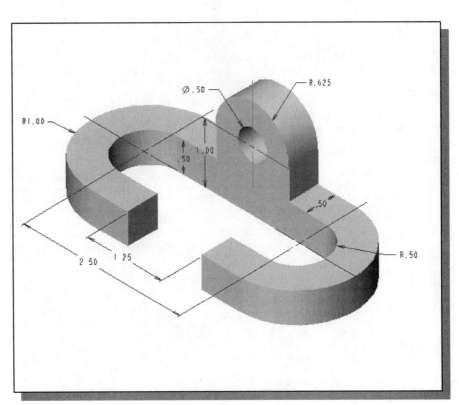

NOTES:

Chapter 4
Model History Tree

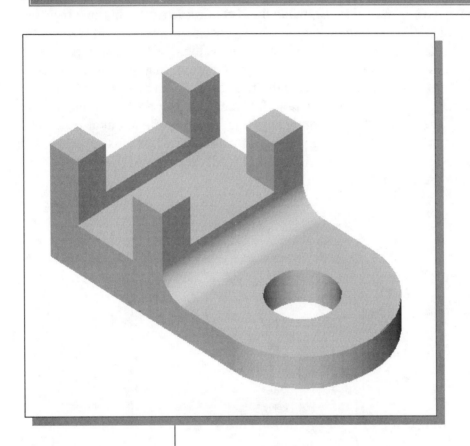

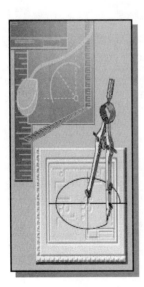

Learning Objectives

♦ **Understand Feature Interactions**
♦ **Use the Part Browser**
♦ **Modify and Update Feature Dimensions**
♦ **Perform History-Based Part Modifications**
♦ **Change the Names of Created Features**
♦ **Implement Basic Design Changes**

Introduction

In *Autodesk Inventor*, the **design intents** are embedded into features in the **history tree**. The structure of the model history tree resembles that of a **CSG binary tree**. A CSG binary tree contains only *Boolean relations*, while the ***Autodesk Inventor* history tree** contains all features, including *Boolean relations*. A history tree is a sequential record of the features used to create the part. This history tree contains the construction steps, plus the rules defining the design intent of each construction operation. In a history tree, each time a new modeling event is created previously defined features can be used to define information such as size, location, and orientation. It is therefore important to think about your modeling strategy before you start creating anything. It is important, but also difficult, to plan ahead for all possible design changes that might occur. This approach in modeling is a major difference of **FEATURE-BASED CAD SOFTWARE**, such as *Autodesk Inventor*, from previous generation CAD systems.

Sequential record of
the construction steps

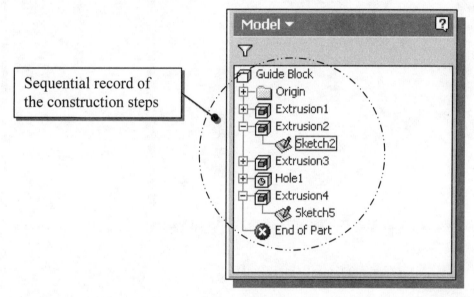

Feature-based parametric modeling is a cumulative process. Every time a new feature is added, a new result is created and the feature is also added to the history tree. The database also includes parameters of features that were used to define them. All of this happens automatically as features are created and manipulated. At this point, it is important to understand that all of this information is retained, and modifications are done based on the same input information.

In *Autodesk Inventor*, the history tree gives information about modeling order and other information about the feature. Part modifications can be accomplished by accessing the features in the history tree. It is therefore important to understand and utilize the feature history tree to modify designs. *Autodesk Inventor* remembers the history of a part, including all the rules that were used to create it, so that changes can be made to any operation that was performed to create the part. In *Autodesk Inventor*, to modify a feature, we access the feature by selecting the feature in the ***Browser*** window.

The *Saddle Bracket* Design

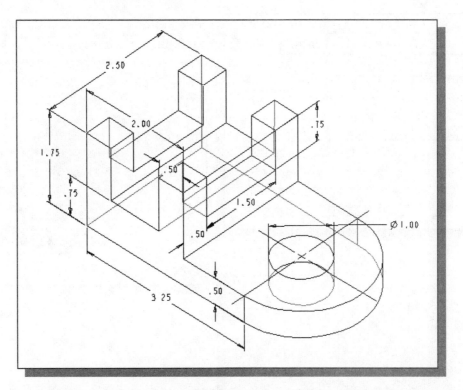

❖ Based on your knowledge of *Autodesk Inventor* so far, how many features would you use to create the design? Which feature would you choose as the **BASE FEATURE**, the first solid feature, of the model? What is your choice in arranging the order of the features? Would you organize the features differently if additional fillets were to be added in the design? Take a few minutes to consider these questions and do preliminary planning by sketching on a piece of paper. You are also encouraged to create the model on your own prior to following through the tutorial.

Starting *Autodesk Inventor*

1. Select the **Autodesk Inventor** option on the *Start* menu or select the **Autodesk Inventor** icon on the desktop to start *Autodesk Inventor*. The *Autodesk Inventor* main window will appear on the screen.

2. Once the program is loaded into memory, the *Startup* dialog box appears at the center of the screen.

3. Select the **New** icon with a single click of the left-mouse-button in the *What to Do* dialog box.

4. Select the **English** tab and in the *New - Choose Template* area, select **Standard(in).ipt**.

5. Pick **OK** in the *Startup* dialog box to accept the selected settings.

Modeling Strategy

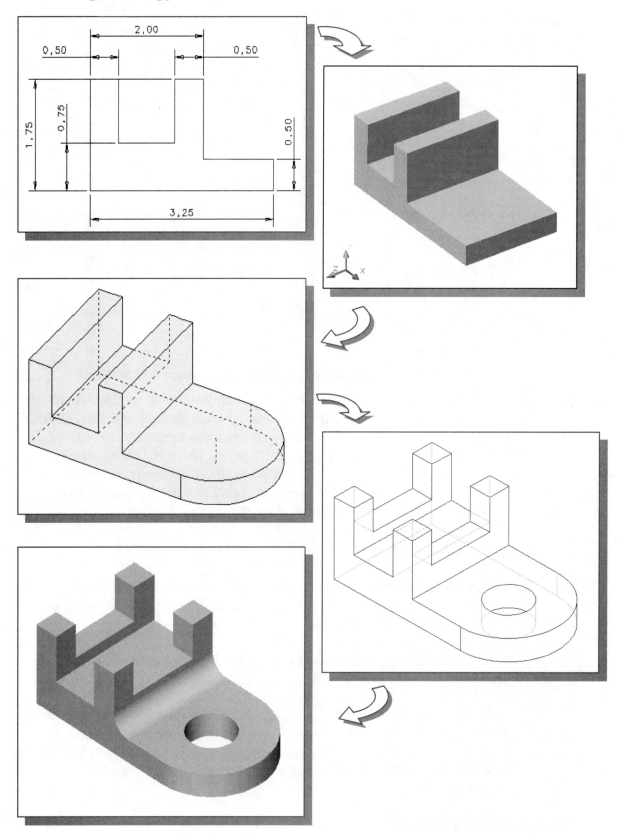

The *Autodesk Inventor Browser*

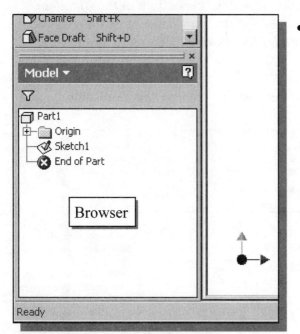

- In the *Autodesk Inventor* screen layout, the **Browser** is located to the left of the graphics window. *Autodesk Inventor* can be used for part modeling, assembly modeling, part drawings, and assembly presentation. The *Browser* window provides a visual structure of the features, constraints, and attributes that are used to create the part, assembly, or scene. The *Browser* also provides right-click menu access for tasks associated specifically with the part or feature, and it is the primary focus for executing many of the *Autodesk Inventor* commands.

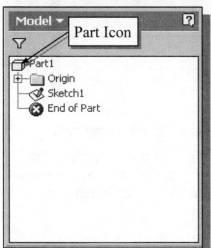

- The first item displayed in the *Browser* is the name of the part, which is also the file name. By default, the name "*Part1*" is used when we first started *Autodesk Inventor*. The *Browser* can also be used to modify parts and assemblies by moving, deleting, or renaming items within the hierarchy. Any changes made in the *Browser* directly affect the part or assembly and the results of the modifications are displayed on screen instantly. The *Browser* also reports any problems and conflicts during the modification and updating procedure.

Creating the Base Feature

1. Move the graphics cursor to the **Line** icon in the *2D Sketch* toolbar. A *Help-tip box* appears next to the cursor and a brief description of the command is displayed at the bottom of the drawing screen: "*Creates Straight line segments and tangent arcs.*" Click once with the left-mouse-button to select the command.

2. Select the icon by clicking once with the left-mouse-button; this will activate the Line command. In the *Status Bar* area, near the bottom of the *Autodesk Inventor* drawing screen, the message "*Specify start point, drag off endpoint for tangent arcs*" is displayed. *Autodesk Inventor* expects us to identify the starting location of a straight line.

3. On your own, create and adjust the geometry by adding and modifying dimensions as shown below.

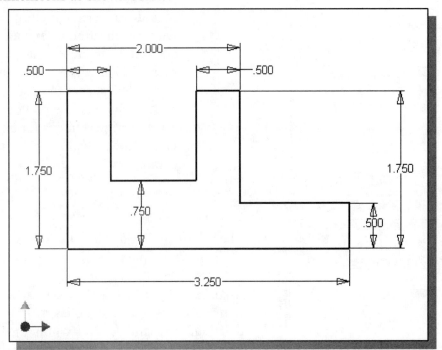

4. Inside the graphics window, click once with the right-mouse-button to display the option menu. Select **Finish Sketch** in the popup menu to end the Sketch option.

5. Press the function key **F6** once to change the display to the *isometric view*.

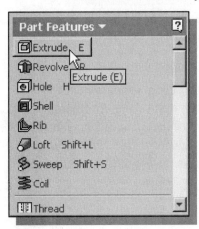

6. In the *Part Feature* toolbar (the toolbar that is located to the left side of the graphics window), select the **Extrude** command by releasing the left-mouse-button on the icon.

7. In the *Distance* option box, enter **2.5** as the total extrusion distance.

8. In the *Extrude* popup window, left-click once on the **Mid-plane** icon. The **Mid-plane** option allows us to extrude in both directions of the sketched profile.

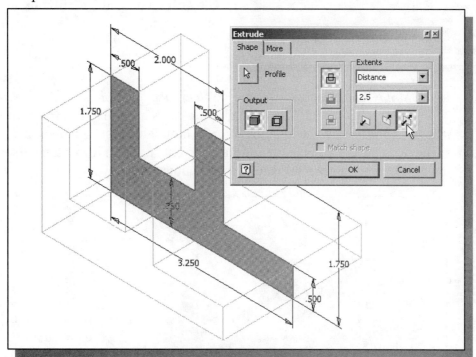

9. Click on the **OK** button to accept the settings and create the base feature.

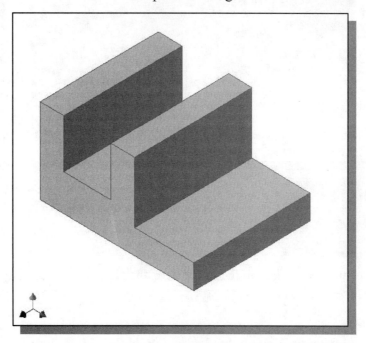

➤ On your own, use the *Dynamic Viewing* functions to view the 3D model. Also notice the extrusion feature is added to the *Model Tree* in the *Browser* area.

Adding the Second Solid Feature

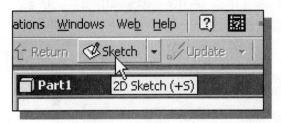

1. In the *Standard* toolbar select the **Sketch** command by left-clicking once on the icon.

2. In the *Status Bar* area, the message: "*Select face, workplane, sketch or sketch geometry.*" is displayed. Move the graphics cursor on the 3D part and notice that *Autodesk Inventor* will automatically highlight feasible planes and surfaces as the cursor is on top of the different surfaces. Move the cursor inside the upper horizontal face of the 3D object as shown below.

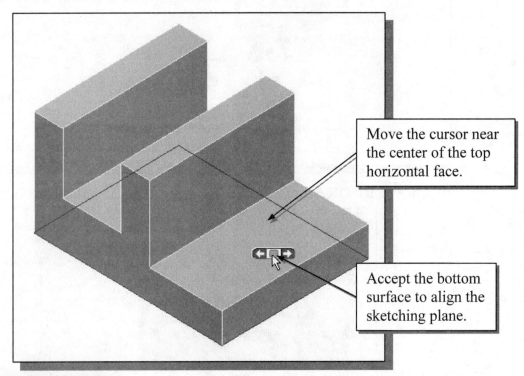

Move the cursor near the center of the top horizontal face.

Accept the bottom surface to align the sketching plane.

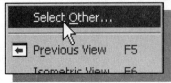

3. Click once with the **right-mouse-button** to bring up the option menu and select **Select Other** to switch to the next feasible choice.

4. On your own, click on the left and/or the right arrows to examine all possible surface selections.

5. Click on the **green** button to select the **bottom horizontal face** of the solid model when it is highlighted as shown in the above figure.

Creating a 2D Sketch

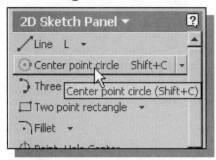

1. Select the **Center point circle** command by clicking once with the left-mouse-button on the icon in the *Sketch* toolbar.

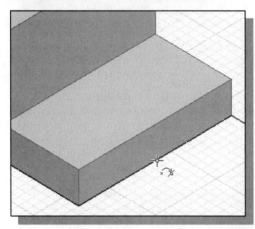

➢ We will align the center of the circle to the midpoint of the base feature.

2. Move the cursor along the shorter edge of the base feature and pick the midpoint of the edge when the midpoint is displayed with a GREEN color as shown in the figure.

3. Select the front corner of the base feature to create a circle as shown below.

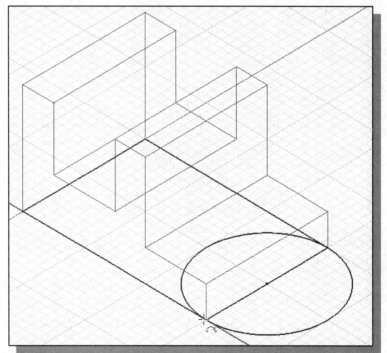

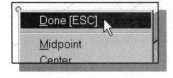

4. Inside the graphics window, click once with the right-mouse-button to display the option menu. Select **Done** in the popup menu to end the Circle command.

5. Inside the graphics window, click once with the right-mouse-button to display the option menu. Select **Finish Sketch** in the popup menu to end the **Sketch** option.

6. In the *Part Features* toolbar (the toolbar that is located to the left side of the graphics window), select the **Extrude** command by releasing the left-mouse-button on the icon.

7. In the *Extrude* popup window, the **Profile** button is pressed down; *Autodesk Inventor* expects us to identify the profile to be extruded. Move the cursor inside the circle we just created and left-click once to select the region as the profile to be extruded.

8. In the *Extrude* popup window, set the *Extents* option to **To** as shown below. Confirm the operation option is set to **Join**.

9. Select the top face of the base feature as the termination surface for the extrusion.

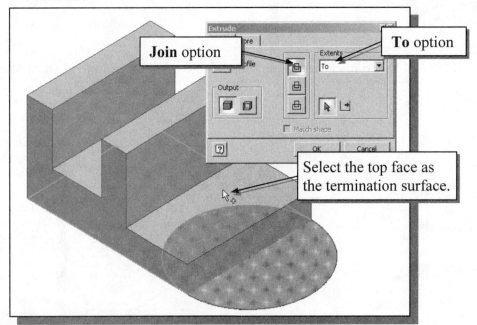

Join option

To option

Select the top face as the termination surface.

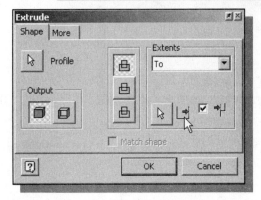

10. Switch *ON* the **option** located in the *Extents* option area; the selected surface will be used as the termination surface.

11. Click on the **OK** button to proceed with the *Join* operation.

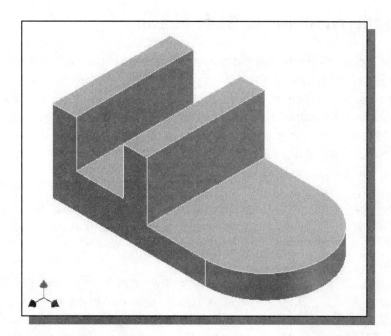

Renaming the Part Features

♦ Currently, our model contains two extruded features. The feature is highlighted in the display area when we select the feature in the *Browser* window. Each time a new feature is created, the feature is also displayed in the *Model Tree* window. By default, *Autodesk Inventor* will use generic names for part features. However, when we begin to deal with parts with a large number of features, it will be much easier to identify the features using more meaningful names. Two methods can be used to rename the features: 1. **Clicking** twice on the name of the feature and 2. Using the **Properties** option. In this example, the use of the first method is illustrated.

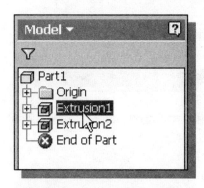

1. Select the first extruded feature in the *Model Browser* area by left-clicking once on the name of the feature, ***Extrusion1***. Notice the selected feature is highlighted in the graphics window.

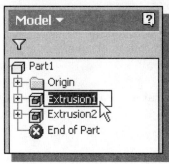

2. Left-mouse-click on the feature name again to enter the Edit mode as shown.

3. Enter ***Base*** as the new name for the first extruded feature.

4. On your own, rename the second extruded feature to ***Circular_End***.

Adjusting the Width of the Base Feature

❖ One of the main advantages of parametric modeling is the ease of performing part modifications at any time in the design process. Part modifications can be done through accessing the features in the history tree. *Autodesk Inventor* remembers the history of a part, including all the rules that were used to create it, so that changes can be made to any operation that was performed to create the part. For our *Saddle Bracket* design, we will reduce the size of the base feature from 3.25 inches to 3.0 inches, and the extrusion distance to 2.0 inches.

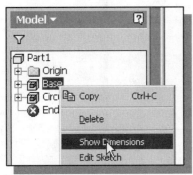

1. Select the first extruded feature, **Base**, in the *Browser* area. Notice the selected feature is highlighted in the graphics window.

2. Inside the *Browser* area, **right-mouse-click** on the first extruded feature to bring up the option menu and select the **Show Dimensions** option in the pop-up menu.

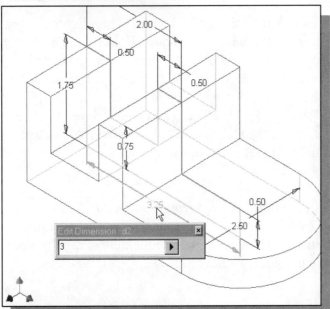

3. All dimensions used to create the **Base** feature are displayed on the screen. Select the overall width of the **Base** feature, the **3.25** dimension value, by double-clicking on the dimension text as shown below.

4. Enter **3.0** in the *Edit Dimension* window.

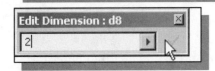

5. On your own, repeat the above steps and modify the extruded distance from **2.5** to **2.0**.

6. Click **Update** in the *Standard* toolbar.

➤ Note that *Autodesk Inventor* updates the model by re-linking all elements used to create the model. Any problems or conflicts that occur will also be displayed during the updating process.

Adding a Placed Feature

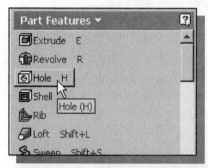

1. In the *Part Features* toolbar (the toolbar that is located to the left side of the graphics window), select the **Hole** command by releasing the left-mouse-button on the icon.

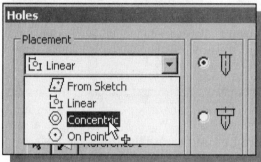

2. In the *Holes* dialog box, choose **Concentric** in the placement option as shown.

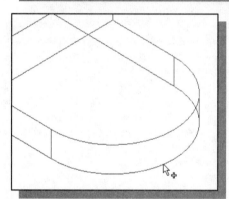

3. Pick the bottom plane of the solid model as the placement plane as shown.

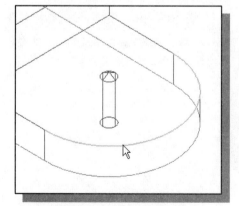

4. Pick the top arc to use as the concentric reference.

5. Set the termination option to **Through All** as shown.

6.　Set the hole diameter to **0.75 in** as shown.

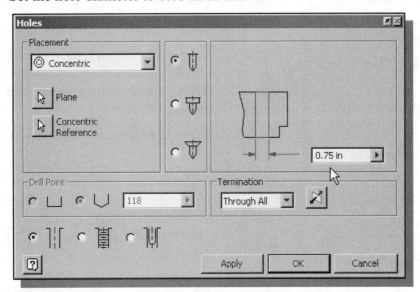

7.　Click **OK** to accept the settings and create the hole feature.

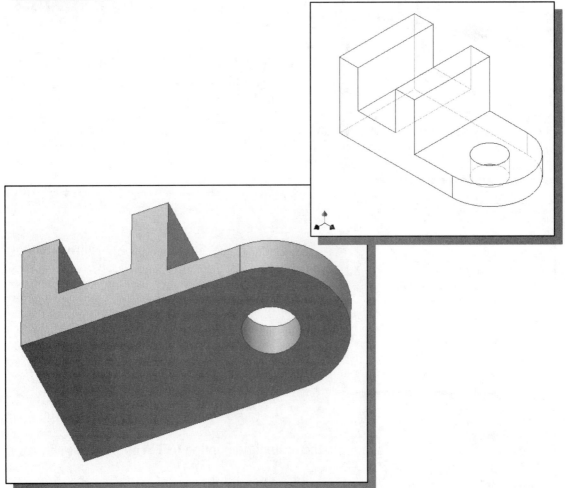

Creating a Rectangular Cut Feature

1. In the *Standard* toolbar select the **Sketch** command by left-clicking once on the icon.

2. Pick the **vertical face** of the solid as shown. (Note the alignment of the origin of the sketch plane.)

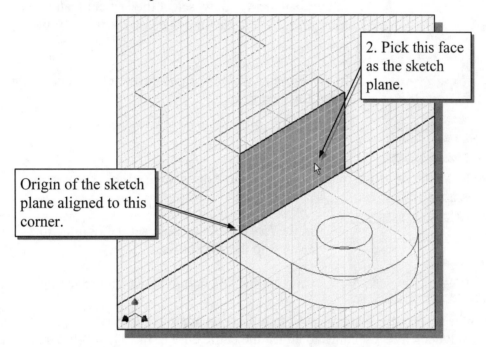

2. Pick this face as the sketch plane.

Origin of the sketch plane aligned to this corner.

➤ On your own, create a rectangular cut (**1.0 x 0.75**) feature (**To Next** option) as shown and rename the feature to *Rect_Cut*.

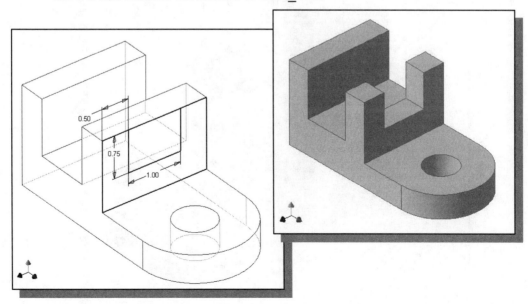

0.50

0.75

1.00

History-based Part Modifications

- *Autodesk Inventor* uses the *history-based part modification* approach, which enables us to make modifications to the appropriate features and re-link the rest of the history tree without having to reconstruct the model from scratch. We can think of it as going back in time and modifying some aspects of the modeling steps used to create the part. We can modify any feature that we have created. As an example, we will adjust the depth of the rectangular cutout.

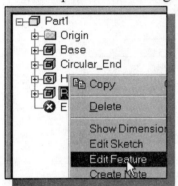

1. In the *Browser* window, select the last cut feature, **Rect_Cut,** by left-clicking once on the name of the feature.

2. In the *Browser* window, right-mouse-click once on the **Rect_Cut** feature.

3. Select **Edit Feature** in the pop-up menu. Notice the *Extrude* dialog box appears on the screen.

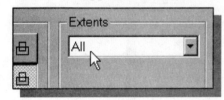

4. In the *Extrude* dialog box, set the termination *Extents* to the **All** option.

5. Click on the **OK** button to accept the settings.

- As can been seen, the history-based modification approach is very straightforward and it only took a few seconds to adjust the cut feature to the **Through All** option.

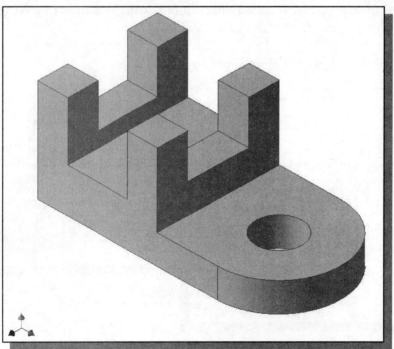

A Design Change

❖ Engineering designs usually go through many revisions and changes. *Autodesk Inventor* provides an assortment of tools to handle design changes quickly and effectively. We will demonstrate some of the tools available by changing the ***Base*** feature of the design.

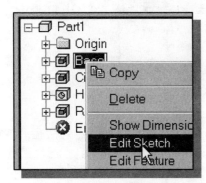

1. In the *Browser* window, select the ***Base*** feature by left-clicking once on the name of the feature.

2. In the *Browser*, right-mouse-click once on the ***Base*** feature to bring up the option menu; then pick **Edit Sketch** in the pop-up menu.

❖ *Autodesk Inventor* will now display the original 2D sketch of the selected feature in the graphics window. We have literally gone back in time to the point where we first created the 2D sketch. Notice the feature being modified is also highlighted in the desktop *Browser*.

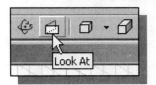

3. Click on the **Look At** icon in the *Standard* toolbar area.

• The **Look At** command automatically aligns the *sketch plane* of a selected entity to the screen.

4. Select any line segment of the 2D sketch. We have literally gone back in time to the point where we first created the 2D sketch.

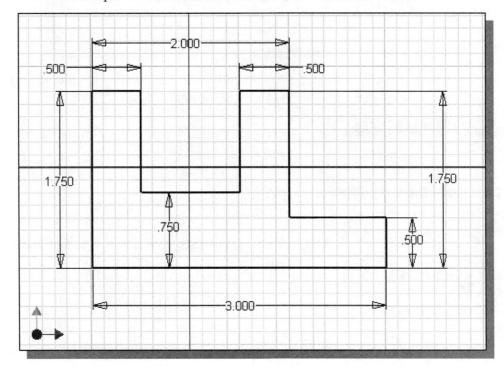

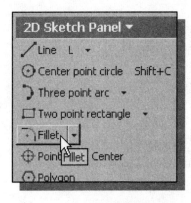

5. Select the **Fillet** command in the *2D Sketch* toolbar.

6. In the graphics window, enter **0.25** as the new radius of the fillet.

7. Select the two edges as shown to create the fillet.

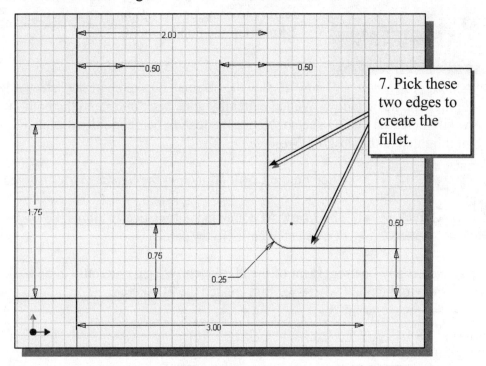

7. Pick these two edges to create the fillet.

- Note that the fillet is created automatically with the dimension attached. The attached dimension can also be modified through the history tree.

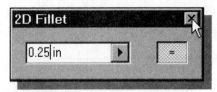

8. Click on the [**X**] icon in the *2D Fillet* window to end the **Fillet** command.

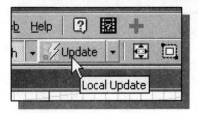

9. Click on the **Update** icon in the *Standard* toolbar area as shown.

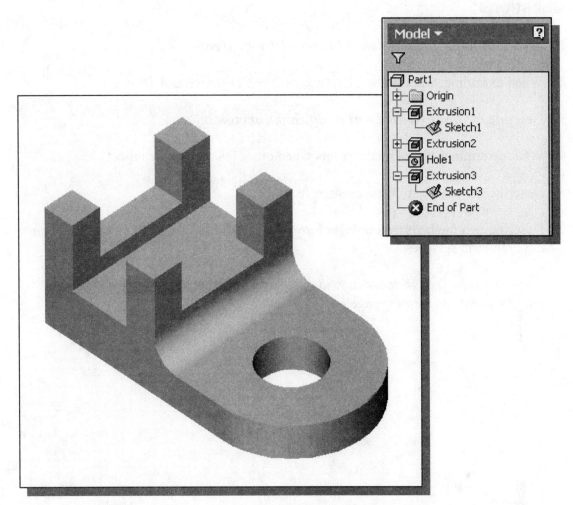

❖ In a typical design process, the initial design will undergo many analyses, testing, and reviews. The *history-based part modification* approach is an extremely powerful tool that enables us to quickly update the design. At the same time, it is quite clear that PLANNING AHEAD is also important in doing feature-based modeling.

Questions:

1. What are stored in the *Autodesk Inventor History Tree*?

2. When extruding, what is the difference between *Distance* and *Through All*?

3. Describe the *history-based part modification* approach.

4. What determines how a model reacts when other features in the model change?

5. Describe the steps to rename existing features.

6. Describe two methods available in *Autodesk Inventor* to *modify the dimension values* of parametric sketches.

7. Create *History Tree sketches* showing the steps you plan to use to create the two models shown on the next page:

Ex.1)

Ex.2)

Exercises: (Dimensions are in inches.)

1. Plate thickness: **0.25** inches.

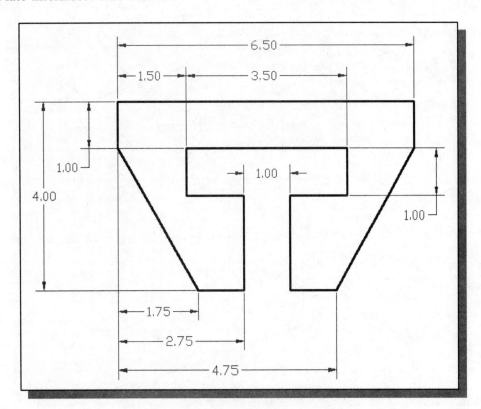

2. Base plate thickness: **0.25** inches. Boss height **0.5** inches.

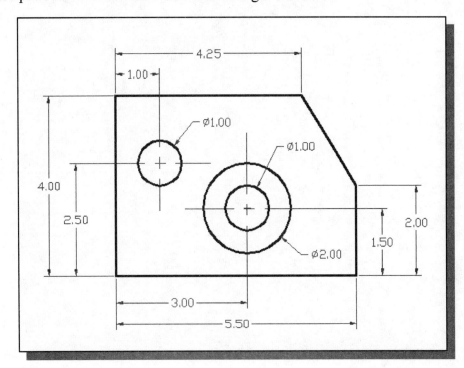

3.

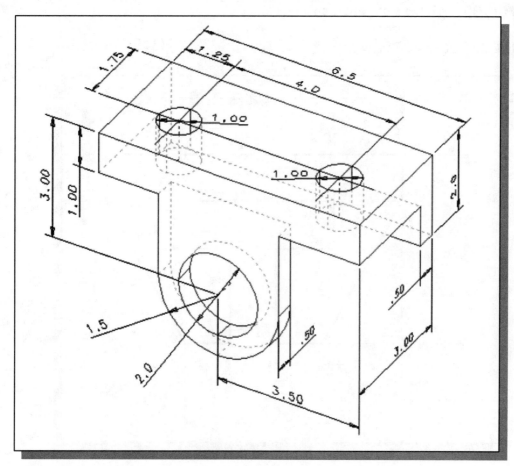

Chapter 5
Parametric Constraints Fundamentals

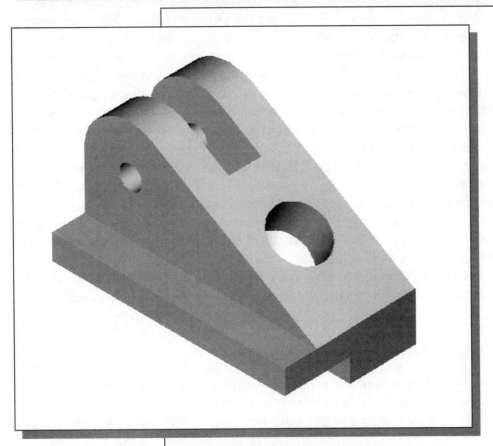

Learning Objectives

♦ **Create Parametric Relations**
♦ **Use Dimensional Variables**
♦ **Display, Add, and Delete Geometric Constraints**
♦ **Understand and Apply Different Geometric Constraints**
♦ **Display and Modify Parametric Relations**
♦ **Create Fully Constrained Sketches**

CONSTRAINTS and RELATIONS

A primary and essential difference between parametric modeling and previous generation computer modeling is that parametric modeling captures the *design intent*. In the previous lessons, we have seen that the design philosophy of *"shape before size"* is implemented through the use of *Autodesk Inventor's* **Profile** and **Dimension** commands. In performing geometric constructions, dimensional values are necessary to describe the **SIZE** and **LOCATION** of constructed geometric entities. Besides using dimensions to define the geometry, we can also apply geometric rules to control geometric entities. More importantly, *Autodesk Inventor* can capture design intent through the use of **geometric constraints**, **dimensional constraints**, and **parametric relations**. In *Autodesk Inventor*, there are two types of constraints: **geometric constraints** and **dimensional constraints**. For part modeling in *Autodesk Inventor*, constraints are applied to *2D sketches*. **Geometric constraints** are **geometric restrictions** that can be applied to geometric entities; for example, *horizontal*, *parallel*, *perpendicular*, and *tangent* are commonly used *geometric constraints* in parametric modeling. **Dimensional constraints** are used to describe the SIZE and LOCATION of individual geometric shapes. In *Autodesk Inventor*, **parametric relations** are user-defined mathematical equations composed of dimensional variables and/or *design variables*. In parametric modeling, features are made of geometric entities with both relations and constraints describing individual design intent. In this lesson, we will discuss the fundamentals of parametric relations and geometric constraints.

Create a *Simple Triangular Plate* Design

➢ In parametric modeling, **geometric properties** such as *horizontal*, *parallel*, *perpendicular*, and *tangent* can be applied to geometric entities automatically or manually. By carefully applying proper **geometric constraints**, very intelligent models can be created. This concept is illustrated by the following example.

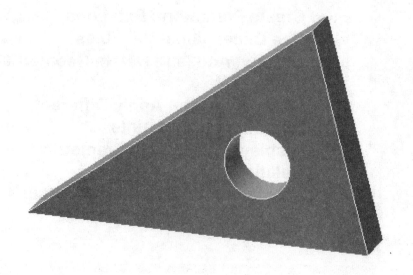

Fully Constrained Geometry

In *Autodesk Inventor*, as we create 2D sketches, geometric constraints such as *horizontal* and *parallel* are automatically added to the sketched geometry. In most cases, additional constraints and dimensions are needed to fully describe the sketched geometry beyond the geometric constraints added by the system. Although we can use *Autodesk Inventor* to build partially constrained or totally unconstrained solid models, the models may behave unpredictably as changes are made. In most cases, it is important to consider the design intent and to add proper constraints to geometric entities. In the following sections, a simple triangle is used to illustrate the different tools that are available in *Autodesk Inventor* to create/modify geometric and dimensional constraints.

Starting Autodesk Inventor

1. Select the **Autodesk Inventor** option on the *Start* menu or select the **Autodesk Inventor** icon on the desktop to start *Autodesk Inventor*. The *Autodesk Inventor* main window will appear on the screen. Once the program is loaded into memory, the *Startup* dialog box appears at the center of the screen.

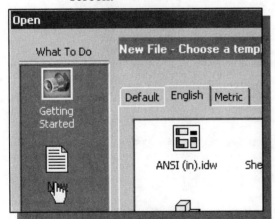

2. Select the **New** icon with a single click of the left-mouse-button in the *What to Do* dialog box.

3. Select the **English** tab and in the *New - Choose a Template* area select **Standard(in).ipt**.

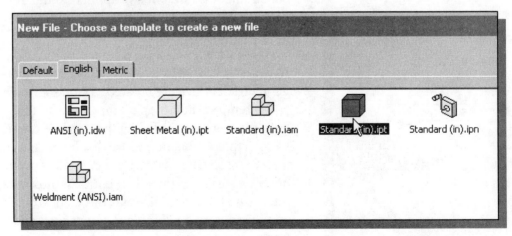

4. Click **OK** in the *Startup* dialog box to accept the selected settings.

5. Move the graphics cursor to the **Line** icon in the *Sketch* toolbar. A *Help-tip box* appears next to the cursor and a brief description of the command is displayed at the bottom of the drawing screen: "*Creates Straight line segments and tangent arcs.*"

6. Create a triangle of arbitrary size positioned near the center of the screen as shown below. (Note that the base of the triangle is horizontal.)

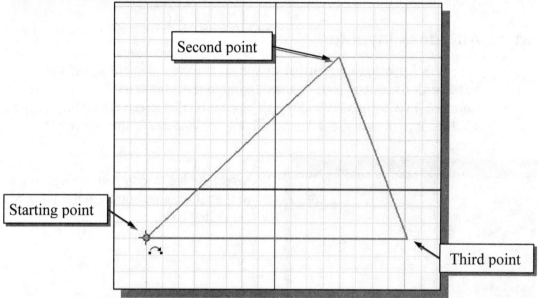

Displaying Existing Constraints

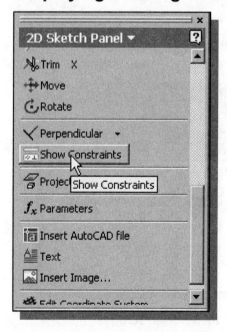

1. Select the **Show Constraints** command in the *Sketch* toolbar. This icon allows us to display constraints that are already applied to the 2D profiles. Left-click once on the icon to display the **Show Constraints**.

➢ In *Autodesk Inventor*, constraints are applied as geometric entities are created. *Autodesk Inventor* will attempt to add proper constraints to the geometric entities based on the way the entities were created. Constraints are displayed as symbols next to the entities as they are created. The current profile consists of three line entities, three straight lines. The horizontal line has three constraints applied to it, two *coincident constraints* and a *horizontal constraint*.

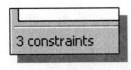

2. Move the cursor on top of the horizontal line and notice the number of constraints applied is displayed in the message area.

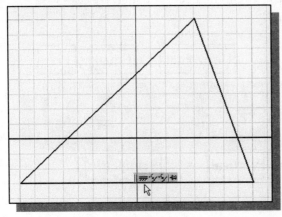

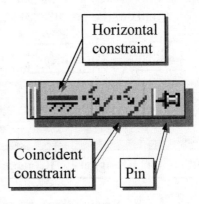

Horizontal constraint

Coincident constraint

Pin

3. Click on the **Pin** icon, the last icon in the *active constraint box*, to lock the display of the *Active Constraints* at its current location.

4. On your own, move the cursor on top of the other two lines and display the active constraint box of the other two entities.

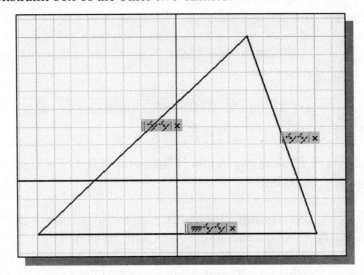

5. Move the cursor on top of the constraints displayed and notice the highlighted endpoint/line indicating the location/entities where the constraints are applied.

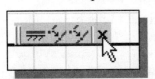

6. Click on the **X** icons, the last icon in the active constraint box, to unlock and turn off the display of the constraints.

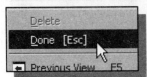

7. Inside the graphics window, right-mouse-click to bring up the option menu and select **Done** to end the Show Constraints command.

Applying Geometric/Dimensional Constraints

- In *Autodesk Inventor*, eleven types of constraints are available. Depending upon the way the constraints are applied, the same results can be accomplished by applying different constraints to the geometric entities.

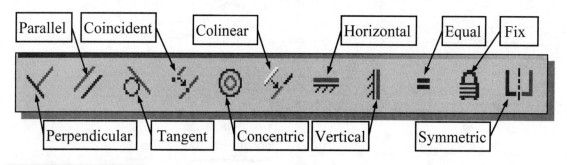

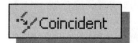 **Perpendicular constraint**: Causes selected curves or ellipse axes to lie at right angles to one another.

 **Parallel constraint**: Causes selected lines or ellipse axes to lie parallel to one another.

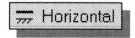 **Tangent constraint**: Constrains two curves to be tangent to one another.

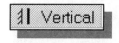 **Coincident constraint**: Constrains two points together or one point to a curve.

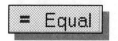 **Concentric constraint**: Constrains two arcs, circles, or ellipses to the same center point.

 Colinear constraint: Causes two lines or ellipse axes to lie along the same line.

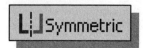 **Horizontal constraint**: Causes lines, ellipse axes, or pairs of points to lie parallel to the X-axis of the sketch coordinate system.

Vertical constraint: Causes lines, ellipse axes, or pairs of points to lie parallel to the Y-axis of the sketch coordinate system.

Equal constraint: Selected arcs/circles constrained to the same radius or selected lines to the same length.

Fixed location constraint: Constrains points or curves to a fixed location relative to the sketch coordinate system.

Symmetric constraint: Constrains lines or curves to become symmetrically constrained about a selected line.

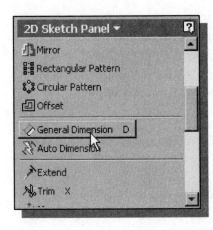

1. Select the **General Dimension** command in the *Sketch* toolbar. The General Dimension command allows us to quickly create and modify dimensions. Left-click once on the icon to activate the General Dimension command.

2. On your own, create the dimension as shown in the figure below. (Note that the displayed value might be different on your screen.)

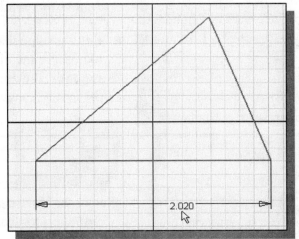

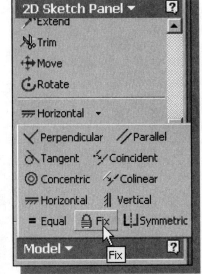

3. Move the cursor to the **Constraint** icon, the icon that is below the Rotate icon, in the *Sketch* toolbar. Click on the triangle icon to display additional options available. A *Help-tip box* appears next to the cursor and a brief description of the command is displayed at the bottom of the drawing screen as the cursor is moved over the different icons.

4. Click on the **Fix** constraint icon to activate the command.

5. Pick the lower right corner of the triangle to make the corner a fixed point.

6. On your own, use the Show Constraints command to confirm the Fix constraint is properly applied.

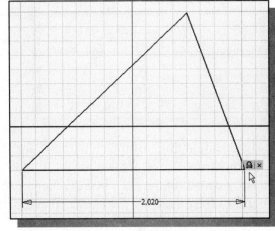

➢ Geometric constraints can be used to control the direction in which changes can occur. For example, in the current design we are adding a horizontal dimension to control the length of the horizontal line. If the length of the line is modified to a greater value, *Autodesk Inventor* will lengthen the line toward the left side. This is due to the fact that the Fix constraint will restrict any horizontal movement of the horizontal line toward the right side

7. Select the **General Dimension** command in the *2D Sketch* toolbar.

8. Click on the dimension text to open the *Edit Dimension* window.

9. Enter a value that is greater than the displayed value to observe the effects of the modification. (For example, the dimension value is 2.02; so enter **3.0** in the text box area.)

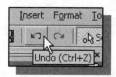

10. On your own, use the **Undo** command to reset the dimension value to the previous value.

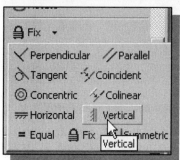

11. Select the **Vertical** constraint icon in the *2D Constraints* toolbar.

12. Pick the inclined line on the right to make the line vertical as shown in the figure below.

13. Hit the **[Esc]** key once to end the Vertical Constraint command.

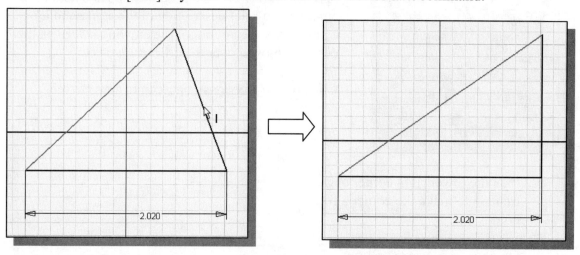

➢ One should think of the constraints and dimensions as defining elements of the geometric entities. How many more constraints or dimensions will be necessary to fully constrain the sketched geometry? Which constraints or dimensions would you use to fully describe the sketched geometry.

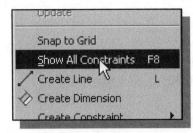

14. Inside the graphics window, click once with the right-mouse-button to display the option menu. Select **Show All Constraints** in the popup menu to show all the applied constraints. (Note that function key **F8** can also be used to activate this command.)

15. Move the cursor on top of the top corner of the triangle.

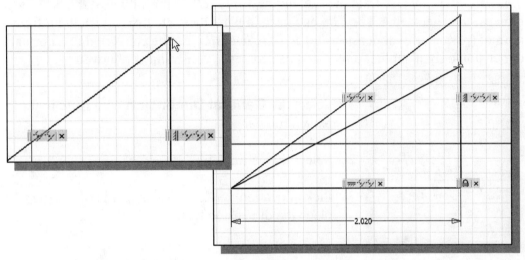

16. Drag the top corner of the triangle and note that the corner can be moved to a new location. Release the mouse button at a new location and notice the corner is adjusted only in an upward or downward direction. Note that the two adjacent lines are automatically adjusted to the new location.

17. On your own, experiment with dragging the other corners to new locations.

- The three constraints that are applied to the geometry provide a full description for the location of the two lower corners of the triangle. The **Vertical** constraint, along with the **Fix** constraint at the lower right corner, does not fully describe the location of the top corner of the triangle. We will need to add additional information, such as the length of the vertical line or an angle dimension.

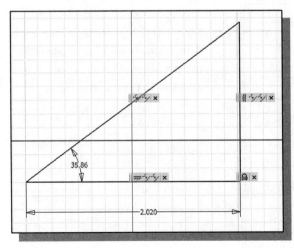

18. On your own, add an angle dimension to the left corner of the triangle.

19. Press the [**Esc**] key once to exit the **General Dimension** command.

20. On your own, modify the angle to **45°** and then experiment with dragging the top corner of the triangle to a new location.

- The geometry is fully constrained with the added dimension.

Over-constraining and Driven Dimensions

- We can use *Autodesk Inventor* to build partially constrained or totally unconstrained solid models. In most cases, these types of models may behave unpredictably as changes are made. However, *Autodesk Inventor* will not let us over-constrain a sketch; additional dimensions can still be added to the sketch, but they are used as references only. These additional dimensions are called **driven dimensions**. *Driven dimensions* do not constrain the sketch; they only reflect the values of the dimensioned geometry. They are enclosed in parentheses to distinguish them from normal (parametric) dimensions. A *driven dimension* can be converted to a normal dimension only if another dimension or geometric constraint is removed.

1. Select the **General Dimension** command in the *Sketch* toolbar.

2. Select the vertical line.

3. Pick a location that is to the right side of the triangle to place the dimension text.

4. A warning dialog box appears on the screen stating that the dimension we are trying to create will over-constrain the sketch. Click on the **Accept** button to proceed with the creation of a driven dimension.

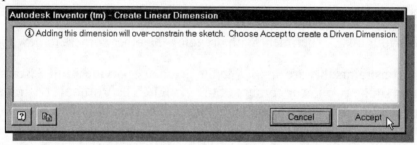

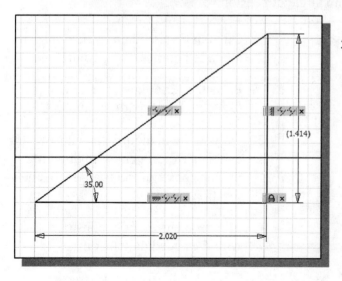

5. On your own, modify the angle dimension to **35°** and observe the changes of the 2D sketch and the driven dimension.

Deleting Existing Constraints

1. On your own, display all the active constraints if they are not already displayed. (Hint: Use the **Show Constraints** command in the *Sketch* toolbar or **Show All Constraints** in the option menu.)

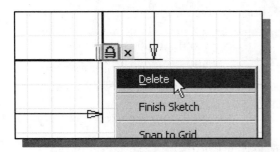

2. Move the cursor on top of the **Fix** constraint icon and right-mouse-click once to bring up the option menu.

3. Select **Delete** to remove the Fix constraint that is applied to the lower right corner of the triangle.

4. Move the cursor on top of the top corner of the triangle.

5. Drag the top corner of the triangle and note that the entire triangle is free to move in all directions. Drag the corner toward the top right corner of the graphics window as shown in the above figure. Release the mouse button to move the triangle to the new location.

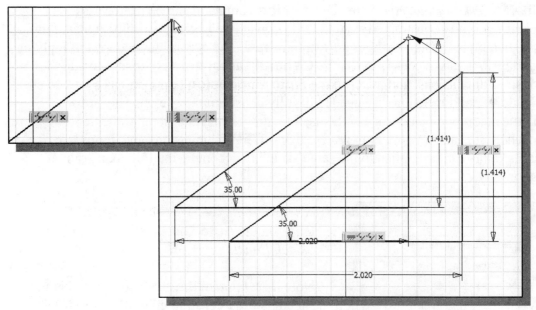

6. On your own, experiment with dragging the other corners and/or the three line segments to new locations on the screen.

❖ **Dimensional constraints** are used to describe the SIZE and LOCATION of individual geometric shapes. **Geometric constraints** are **geometric restrictions** that can be applied to geometric entities. The constraints applied to the triangle are sufficient to maintain its size and shape, but the geometry can be moved around; its location definition isn't complete.

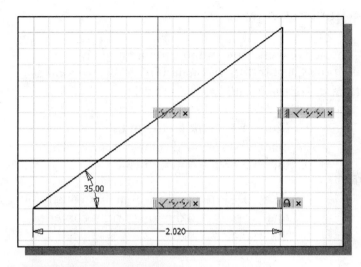

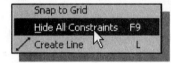

7. On your own, reapply the **Fix** constraint to the lower right corner of the triangle and delete the reference dimension.

8. Confirm the same constraints and dimensions are applied on your sketch as shown.

❖   Note that the sketch is fully constrained.

9. On your own, hit the function key **F9** once to hide all the constraints.

Using the Auto Dimension Command

➢   In *Autodesk Inventor*, the **Auto Dimension** command can be used to assist in creating a fully constrained sketch. Fully constrained sketches can be updated more predictably as design changes are implemented. The general procedure for applying dimensions to sketches is to use the **General Dimension** command to add the desired dimensions, and then use the **Auto Dimension** command as a way to check if additional dimensions are needed to fully constrain the sketch. The Auto Dimension command can also be used to apply some additional dimensions that are needed. It is also important to realize that different dimensions and geometric constraints can be applied to the same sketch to accomplish a fully constrained geometry.

1. Click on the **Auto Dimension** icon in the *2D Sketch Panel*.

❖   Note that *Autodesk Inventor* confirms that the sketch is fully constrained with the message "*0 Dimensions Required.*"

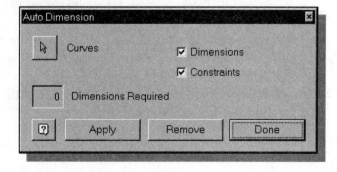

2. Click **Done** to exit the **Auto Dimension** command.

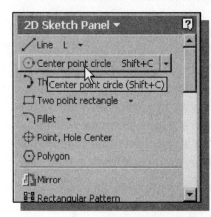

3. Select the **Center point circle** command by clicking once with the left-mouse-button on the icon in the *Sketch* toolbar.

4. On your own, create a circle of arbitrary size inside the triangle as shown below.

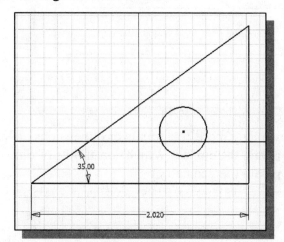

5. Click on the **Auto Dimension** icon in the *2D Sketch Panel*.

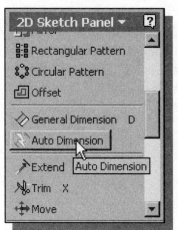

❖ Note that *Autodesk Inventor* confirms that the sketch is not fully constrained and "*3 Dimensions Required*" to fully constrain the circle. What are the dimensions and/or constraints can be applied to fully constrain the circle?

6. Click **Done** to exit the **Auto Dimension** command.

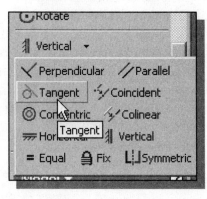

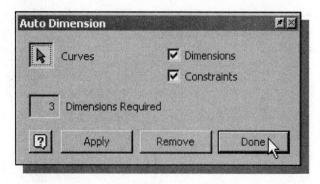

7. Click on the **Tangent** constraint icon in the *Sketch* toolbar.

8. Pick the circle by left-mouse-clicking once on the geometry.

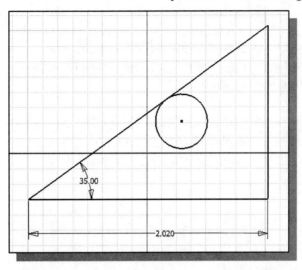

9. Pick the inclined line. The sketched geometry is adjusted as shown.

10. Inside the graphics window, click once with the right-mouse-button to display the option menu. Select **Done** in the popup menu to end the **Tangent** command.

*   How many more constraints or dimensions do you think will be necessary to fully constrain the circle? Which constraints or dimensions would you use to fully constrain the geometry?

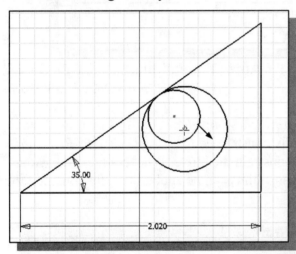

11. Move the cursor on top of the right side of the circle, and then drag the circle toward the right edge of the graphics window. Notice the size of the circle is adjusted while the system maintains the **Tangent** constraint.

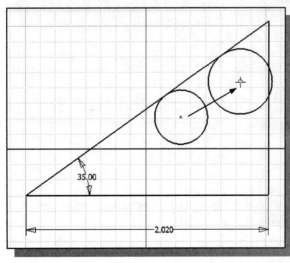

12. Drag the center of the circle toward the upper right direction. Notice the **Tangent** constraint is always maintained by the system.

➢   On your own, experiment with adding additional constraints and/or dimensions to fully constrain the sketched geometry. Use the **Undo** command to undo any changes before proceeding to the next section.

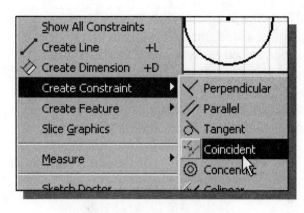

13. Inside the graphics window, click once with the *right-mouse-button* to display the option menu. Select **Create Constraint → Coincident** in the popup menus.

- The option menu is a quick way to access many of the commonly used commands in *Autodesk Inventor*.

14. Pick the vertical line.

15. Pick the center of the circle to align the center of the circle and the vertical line.

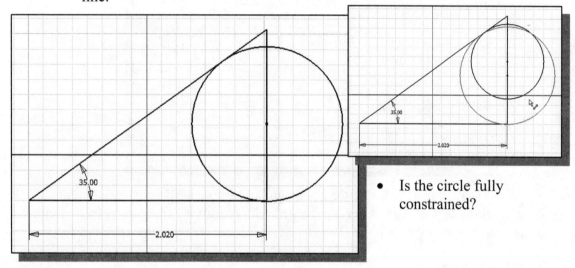

- Is the circle fully constrained?

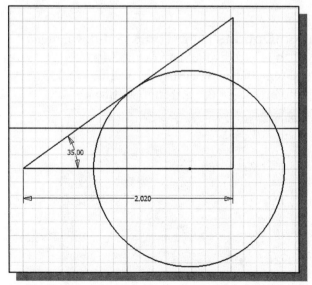

16. On your own, delete the **Coincident** constraint we just applied and add a **Coincident** constraint in between the center of the circle and the horizontal line.

- How many more constraints or dimensions do you think will be necessary to fully constrain the circle? Which constraints or dimensions would you use to fully constrain the geometry?

❖ The application of different constraints affects the geometry differently. The design intent is maintained in the CAD model's database and thus allows us to create very intelligent CAD models that can be modified/revised fairly easily. On your own, experiment and observe the results of applying different constraints to the triangle. For example: (1) adding another **Fix** constraint to the top corner of the triangle; (2) deleting the horizontal dimension and adding another **Fix** constraint to the left corner of the triangle; and (3) adding another **Tangent** constraint and adding the size dimension to the circle.

17. On your own, modify the 2D sketch as shown below.

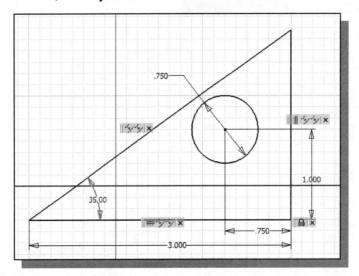

➢ On your own, use the **Extrude** command and create a 3D solid model with a plate thickness of **0.25**. Also experiment with modifying the parametric relations and dimensions through the *Part Browser*.

Constraint Settings

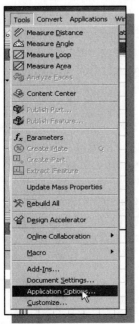

• Select **Application Options** in the **Tools** pull-down menu. Click on the **Sketch** tab to display and/or modify the constraint settings. On your own, adjust the settings and experiment with the effects of the different settings.

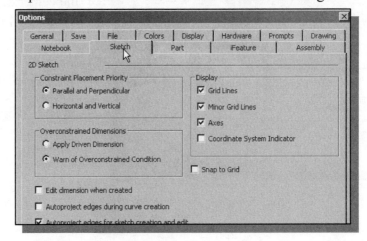

Parametric Relations

- In parametric modeling, dimensions are design parameters that are used to control the sizes and locations of geometric features. Dimensions are more than just values; they can also be used as feature control variables. This concept is illustrated by the following example.

1. Start a new drawing by left-clicking once on the **New →
Part** icon in the *Standard* toolbar.

- Another graphics window appears on the screen. We can switch between the two models by clicking on the different graphics windows.

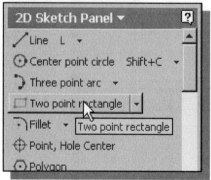

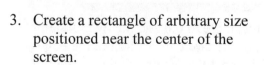

2. Select the **Two point rectangle** command by clicking once with the left-mouse-button on the icon in the *Sketch* toolbar.

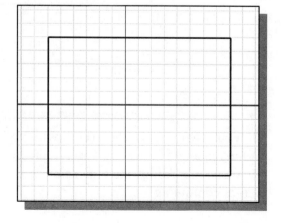

3. Create a rectangle of arbitrary size positioned near the center of the screen.

4. Select the **Center point circle** command by clicking once with the left-mouse-button on the icon in the *Sketch* toolbar.

5. Create a circle of arbitrary size inside the rectangle as shown below.

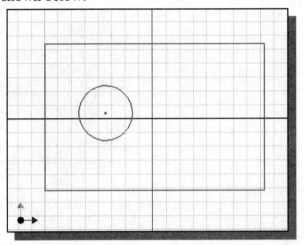

6. On your own, adjust the geometry by adding and modifying dimensions as shown below. (Note: Create and modify the overall width and height of the rectangle first. These two dimensions will be used as the control variables for the rest of the dimensions.)

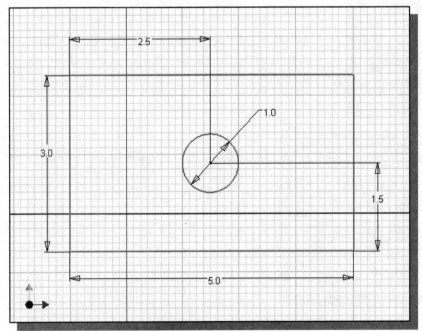

- On your own, change the overall width of the rectangle to **6.0** and the overall height of the rectangle to **3.6** and observe the location of the circle in relation to the edges of the rectangle. Adjust the dimensions back to **5.0** and **3.0** as shown in the above figure before continuing.

Dimensional Values and Dimensional Variables

Initially in *Autodesk Inventor*, values are used to create different geometric entities. The text created by the Dimension command also reflects the actual location or size of the entity. Each dimension is also assigned a name that allows the dimension to be used as a control variable. The default format is "dxx," where the "xx" is a number that *Autodesk Inventor* increments automatically each time a new dimension is added.

Let us look at our current design, which represents a plate with a hole at the center. The dimensional values describe the size and/or location of the plate and the hole. If a modification is required to change the width of the plate, the location of the hole will remain the same as described by the two location dimensional values. This is okay if that is the design intent. On the other hand, the *design intent* may require (1) keeping the hole at the center of the plate and (2) maintaining the size of the hole to be one-third of the height of the plate. We will establish a set of parametric relations using the dimensional variables to capture the design intent described in statements (1) and (2) above.

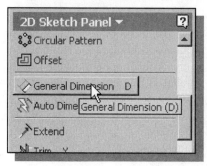

1. Move the cursor on top of the **General Dimension** icon. The General Dimension command allows us to quickly create and modify dimensions. Left-click once on the icon to activate the General Dimension command.

2. Click on the **width dimension** of the rectangle to display the *Edit Dimension* window.

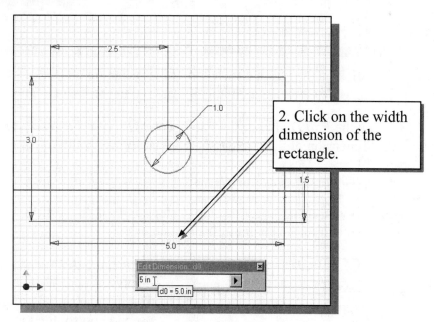

2. Click on the width dimension of the rectangle.

-   Notice the *variable name* d0 is displayed in the title area of the *Edit Dimension* window and also in the cursor box when the cursor is moved near the text box.

3. Click on the *check mark* button to close the *Edit Dimension* window.

Parametric Equations

- Each time we add a dimension to a model, that value is established as a parameter for the model. We can use parameters in equations to set the values of other parameters.

    1. Click on the **horizontal location dimension** of the circle to display the *Edit Dimension* window.

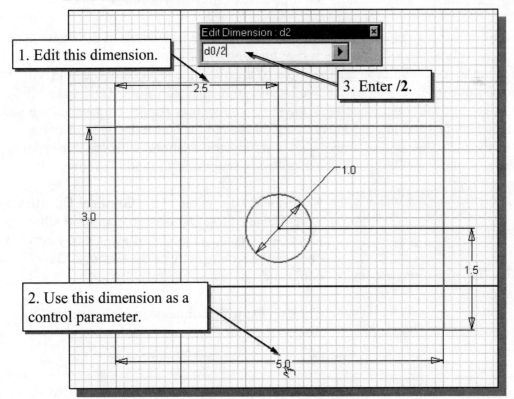

    2. Click on the width dimension of the rectangle **d0** (value of **5.0**). Notice the variable name is automatically entered in the *Edit Dimension* window.

    3. In the *Edit Dimension* window, enter **/2** to set the horizontal location dimension of the circle to be one-half of the width of the rectangle.

    4. Click on the **check mark** button to close the *Edit Dimension* window.

    5. On your own, repeat the above steps and set the vertical location dimension to one-half of the height of the rectangle.

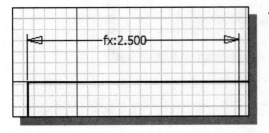

❖ Notice, the derived dimension values are displayed with **fx** in front of the numbers. The parametric relations we entered are used to control the location of the circle; the location is based on the height and width of the rectangle.

Viewing the Established Parameters and Relations

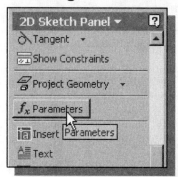

1. In the *2D Sketch* toolbar select the **Parameters** command by left-clicking once on the icon. The *Parameters* popup window appears.

- The **Parameters** command can be used to display all dimensions used to define the model. We can also create additional parameters as design variables, which are called ***user parameters***.

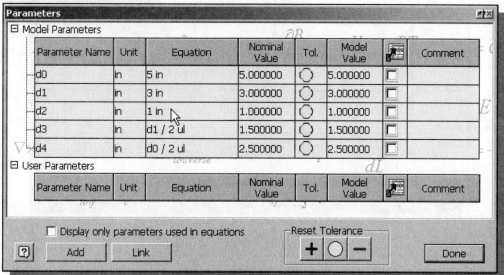

2. Click on the **1.0** value in the equation section of the *Parameters* window. Enter **d1/3** as the parametric relation to set the size of the circle to be one-third of the height of the rectangle as shown below.

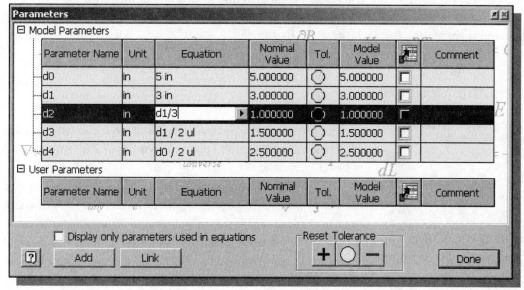

3. Click on the **Done** button to accept the settings.

4. On your own, change the dimensions of the rectangle to **6.0 x 3.6** and observe the changes to the location and size of the circle. (Hint: Double-click the dimension text to bring up the *Edit Dimension* window.)

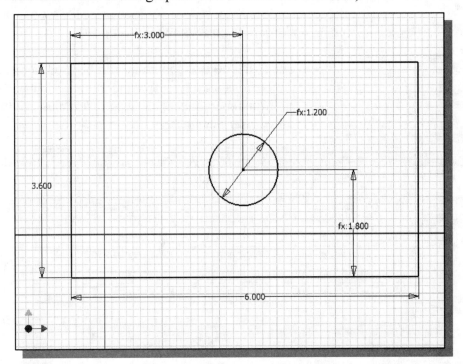

❖ *Autodesk Inventor* automatically adjusts the dimensions of the design, and the parametric relations we entered are also applied and maintained. The dimensional constraints are used to control the size and location of the hole. The design intent, previously expressed by statements (1) and (2) at the beginning of this section, is now embedded into the model.

➤ On your own, use the **Extrude** command and create a 3D solid model with a plate thickness of **0.25**. Also, experiment with modifying the parametric relations and dimensions through the *Part Browser*.

Saving the Model File

1. Select **Save** in the *Standard* toolbar. We can also use the "**Ctrl-S**" combination (press down the "Ctrl" key and hit the "S" key once) to save the part.

2. In the popup window, enter **Plate** as the name of the file.

3. Click on the **SAVE** button to save the file.

Questions:

1. What is the difference between *dimensional* constraints and *geometric* constraints?

2. How can we confirm that a sketch is fully constrained?

3. How do we distinguish between derived dimensions and regular dimensions on the screen?

4. Describe the procedure to Display/Edit user-defined equations.

5. List and describe three different geometric constraints available in *Autodesk Inventor*.

6. Does *Autodesk Inventor* allow us to build partially constrained or totally unconstrained solid models? What are the advantages and disadvantages of building these types of models?

7. Identify and describe the following commands.

(a)

(b)

(c)

(d)

Exercises:

(Create and establish three parametric relations for each of the following designs.)

1. Dimensions are in millimeters. (Base thickness: 10 mm., Boss height: 20 mm.)

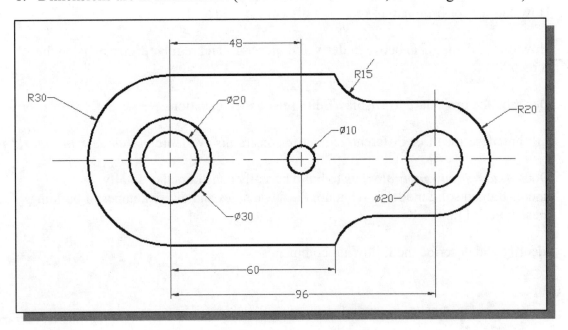

2. Dimensions are in inches. Plate thickness: 0.25 inches.

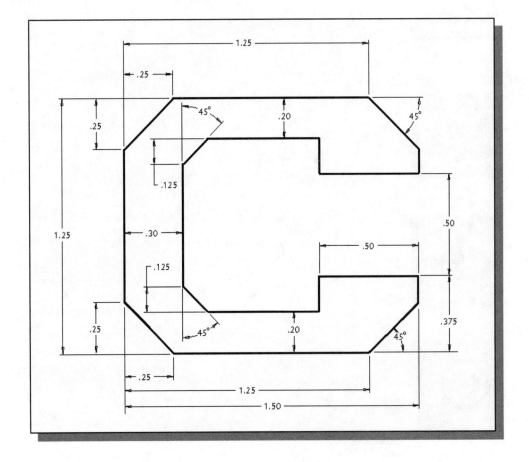

3. Dimensions are in inches

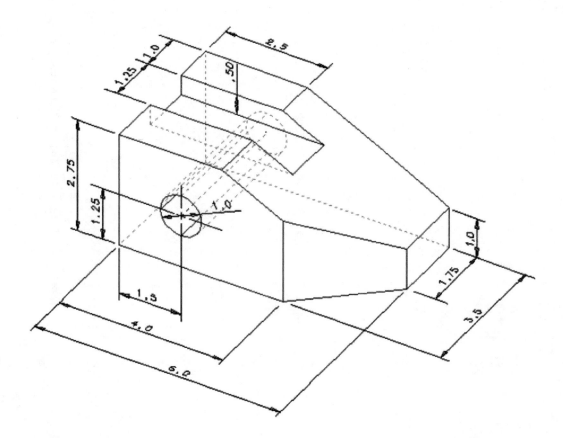

4. Dimensions are in inches.

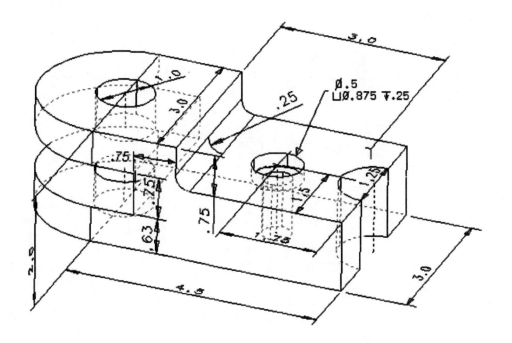

NOTES:

Chapter 6
Geometric Construction Tools

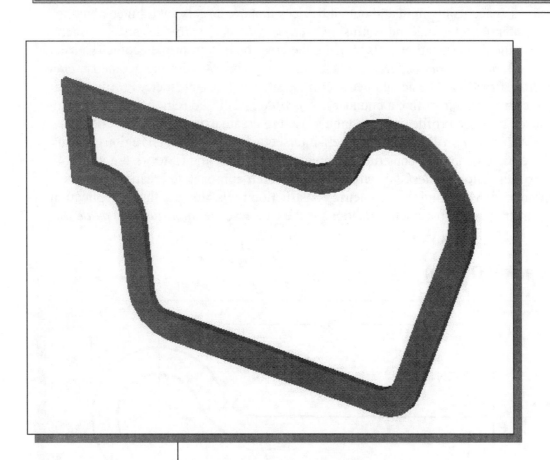

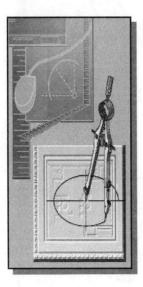

Learning Objectives

♦ **Applying Geometry Constraints**
♦ **Use the Trim/Extend Command**
♦ **Use the Offset Command**
♦ **Understand the Profile Sketch Approach**
♦ **Create Projected Geometry**
♦ **Understanding and Using Reference Geometry**
♦ **Edit with Click and Drag**
♦ **Using the Auto Dimension Command**

Introduction

The main characteristics of solid modeling are the accuracy and completeness of the geometric database of the three-dimensional objects. However, working in three-dimensional space using input and output devices that are largely two-dimensional in nature is potentially tedious and confusing. *Autodesk Inventor* provides an assortment of two-dimensional construction tools to make the creation of wireframe geometry easier and more efficient. *Autodesk Inventor* includes two types of wireframe geometry: **curves** and **profiles**. Curves are basic geometric entities such as lines, arcs, etc. Profiles are a group of curves used to define a boundary. A *profile* is a closed region and can contain other closed regions. Profiles are commonly used to create extruded and revolved features. An *invalid profile* consists of self-intersecting curves or open regions. In this lesson, the basic geometric construction tools, such as Trim and Extend, are used to create profiles. The *Autodesk Inventor's* **profile sketch** approach to creating profiles is also introduced. Mastering the geometric construction tools along with the application of proper constraints and parametric relations is the true essence of *parametric modeling*.

The *Gasket* Design

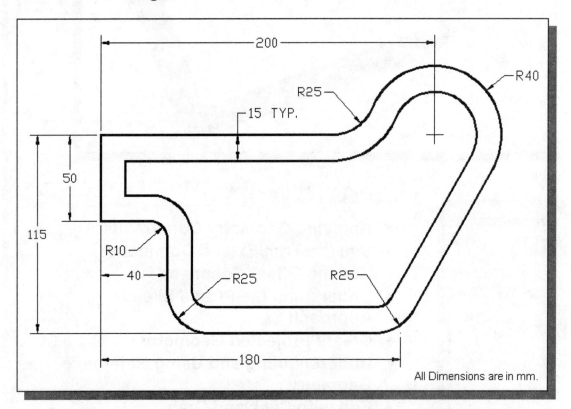

❖   Based on your knowledge of *Autodesk Inventor* so far, how would you create this design? What are the more difficult geometry involved in the design? Take a few minutes to consider a modeling strategy and do preliminary planning by sketching on a piece of paper. You are also encouraged to create the design on your own prior to following through the tutorial.

Modeling Strategy

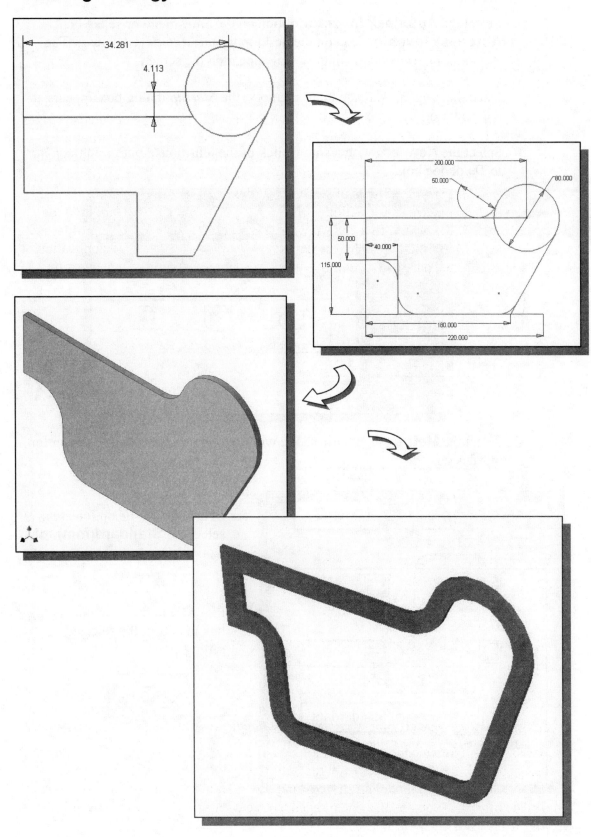

Starting Autodesk Inventor

1. Select the **Autodesk Inventor** option on the *Start* menu or select the **Autodesk Inventor** icon on the desktop to start *Autodesk Inventor*. The *Autodesk Inventor* main window will appear on the screen.

2. Once the program is loaded into memory, the ***Startup*** dialog box appears at the center of the screen.

3. Select the **New** icon with a single click of the left-mouse-button in the *What to Do* dialog box.

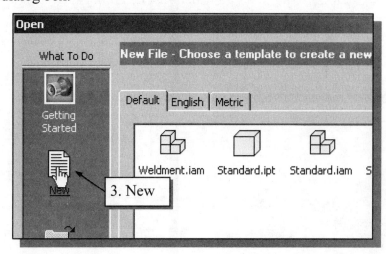

4. Select the **Metric** tab as shown. We will use the millimeter (mm) setting for this design.

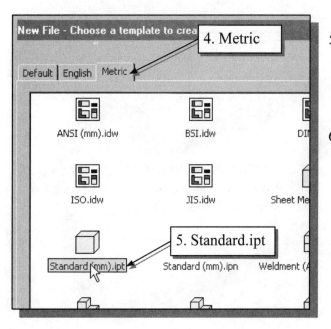

5. In the *New - File* option area, select the **Standard(mm).ipt** icon as shown.

6. Pick **OK** in the *New File* dialog box to accept the selected settings.

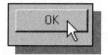

Creating a 2D Sketch

1. Click on the **Line** icon in the *Sketch* toolbar.

2. Create a sketch as shown in the figure below. Start the sketch from the top right corner. The line segments are all parallel and/or perpendicular to each other. We will intentionally make the line segments of arbitrary length, as it is quite common during the design stage that not all of the values are determined.

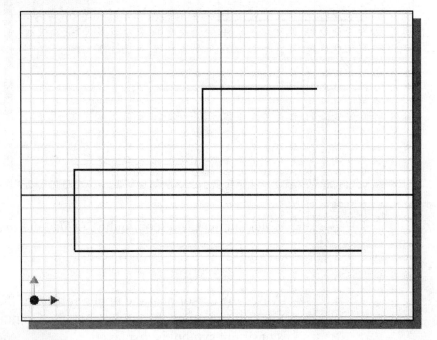

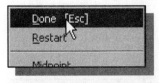

3. Inside the graphics window, right-mouse-click to bring up the option menu and select **Done** to end the Line command.

4. Select the **Center point circle** command by clicking once with the left-mouse-button on the icon in the *2D Sketch Panel*.

5. Pick a location that is above the bottom horizontal line as the center location of the circle.

6. Move the cursor toward the right and create a circle of arbitrary size, by clicking once with the left-mouse-button.

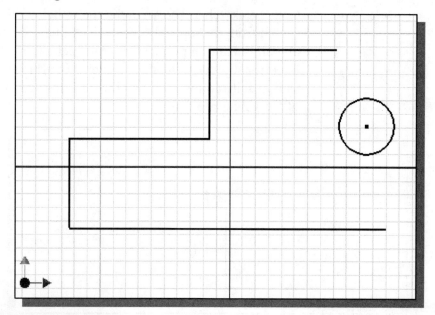

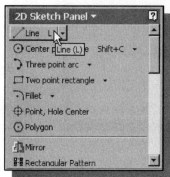

7. Click on the **Line** icon in the *2D Sketch Panel*.

8. Move the cursor near the upper portion of the circle and pick a location on the circle when the Coincident constraint symbol is displayed.

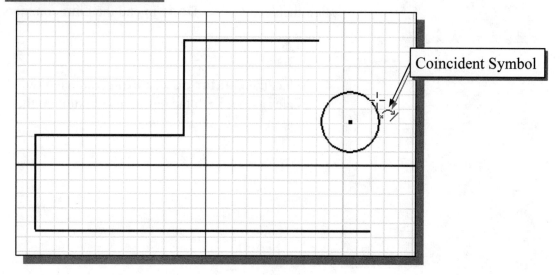

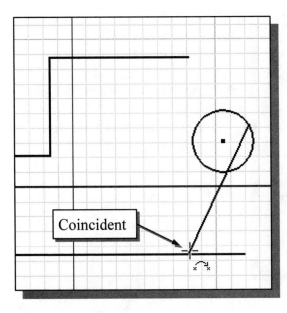

Coincident

9. For the other end of the line, select a location that is on the lower horizontal line and about one-third from the right endpoint. Notice the **Coincident** constraint symbol is displayed when the cursor is on the horizontal line.

10. Inside the graphics window, right-mouse-click to bring up the option menu.

11. In the option menu, select **Done** to end the Line command.

Editing the Sketch by Dragging the Entities

❖ In *Autodesk Inventor*, we can click and drag any under-constrained curve or point in the sketch to change the size or shape of the sketched profile. As illustrated in the previous chapter, this option can be used to identify under-constrained entities. This *Editing by Dragging* method is also an effective visual approach that allows designers to quickly make changes.

1. Move the cursor on the lower left vertical edge of the sketch. Click and drag the edge to a new location that is toward the right side of the sketch.

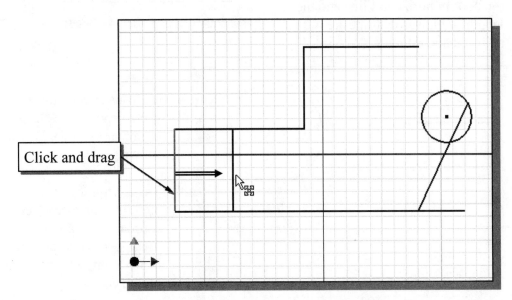

Click and drag

❖ Note that we can only drag the vertical edge horizontally; the connections to the two horizontal lines are maintained while we are moving the geometry.

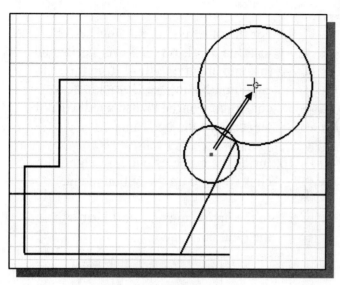

2. Click and drag the center point of the circle to a new location.

❖ Note that as we adjust the size and the location of the circle, the connection to the inclined line is maintained.

3. Click and drag the lower endpoint of the inclined line to a new location.

❖ Note that as we adjust the size and the location of the inclined line the location of the bottom horizontal edge is also adjusted.

➤ Note that several changes occur as we adjust the size and the location of the inclined line. The location of the bottom horizontal line and the length of the vertical line are adjusted accordingly.

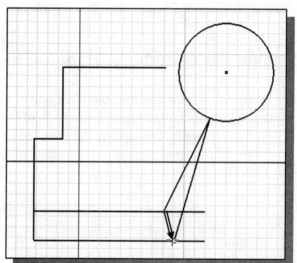

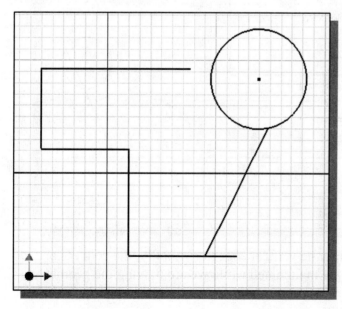

4. On your own, adjust the sketch so that the shape of the sketch appears roughly as shown.

❖ The *Editing by Dragging* method is an effective approach that allows designers to explore and experiment with different design concepts.

Adding Additional Constraints

1. Choose **General Dimension** in the *2D Sketch Panel*.

2. Add the horizontal location dimension, from the top left vertical edge to the center of the circle as shown. (Do not be overly concerned with the dimensional value; we are still working on creating a *rough sketch*.)

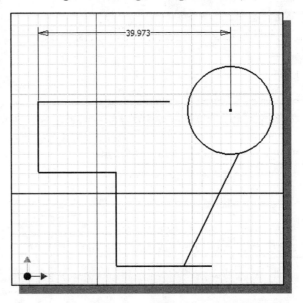

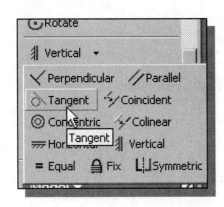

3. Click on the **Tangent** constraint icon in the *2D Sketch Panel*.

4. Pick the inclined line by left-mouse-clicking once on the geometry.

5. Pick the circle. The sketched geometry is adjusted as shown below.

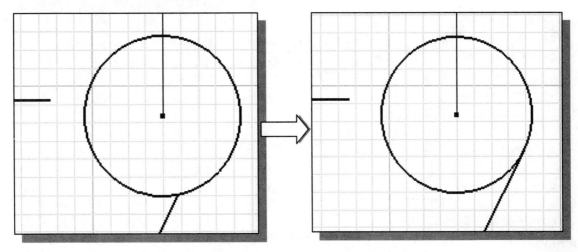

6. Inside the graphics window, right-mouse-click to bring up the option menu.

7. In the option menu, select **Done** to end the Tangent command.

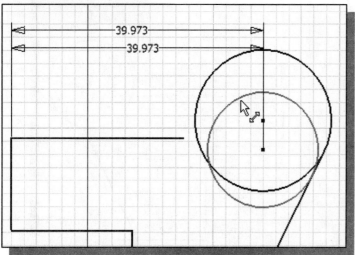

8. Click and drag the circle to a new location.

❖   Note that the dimension we added now restricts the horizontal movement of the center of the circle. The tangent relation to the inclined line is maintained.

Using the Trim and Extend Commands

➢   In the following sections, we will illustrate using the Trim and Extend commands to complete the desired 2D profile.

❖   The **Trim** and **Extend** commands can be used to shorten/lengthen an object so that it ends precisely at a boundary. As a general rule, *Autodesk Inventor* will try to clean up sketches by forming a closed region sketch. Also note that while we are in either **Trim** or **Extend**, we can press the [**Shift**] key to switch to the opposite operation.

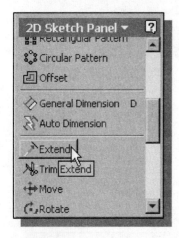

1. Choose **Extend** in the *2D Sketch Panel*. The message "*Extend curves*" is displayed in the prompt area.

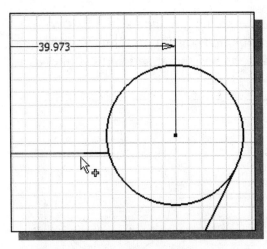

2. We will first extend the top horizontal line to the circle. Move the cursor near the right hand endpoint of the top horizontal line. *Autodesk Inventor* will automatically display the possible result of the selection.

3. We will next trim the bottom horizontal line to the inclined line. The **Trim** operation can be activated by selecting the Trim icon in the *2D Sketch Panel* or by pressing down the [**Shift**] key while we are in the Extend command. Move the cursor near the right hand endpoint of the bottom horizontal line and select the line and press down the [**Shift**] key. *Autodesk Inventor* will display a dashed line indicating the portion of the line that will be trimmed.

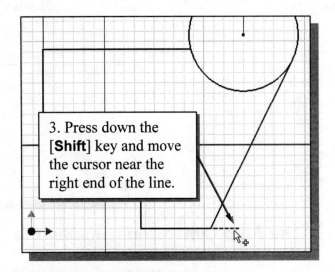

3. Press down the [**Shift**] key and move the cursor near the right end of the line.

4. Left-mouse-click once on the line to perform the Trim operation.

5. On your own, create the vertical location dimension as shown below.

6. Adjust the dimension to **0.0** so that the horizontal line and the center of the circle are aligned horizontally.

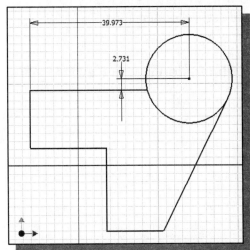

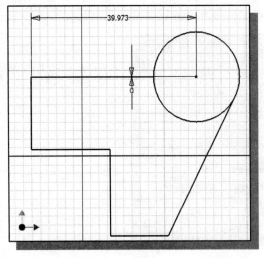

7. On your own, create the **height dimension** to the left and use the **Show Constraints** command to examine the applied constraints. Confirm that a **Perpendicular constraint** is applied to the horizontal and vertical lines as shown.

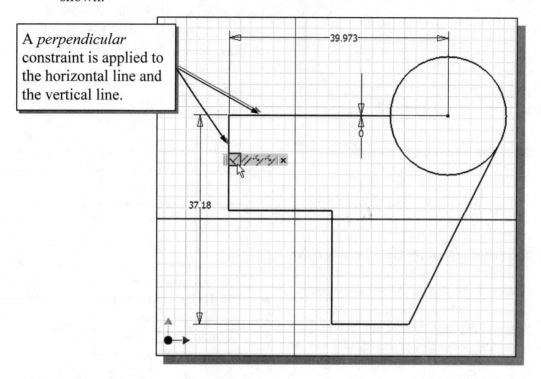

A *perpendicular* constraint is applied to the horizontal line and the vertical line.

The Auto Dimension command

➢ In *Autodesk Inventor*, we can use the **Auto Dimension** command to assist in creating a fully constrained sketch. Fully constrained sketches can be updated more predictably as design changes are implemented. The general procedure for applying dimensions to sketches is to use the **General Dimension** command to add the desired key dimensions, and then use the **Auto Dimension** command as a quick way to calculate all other sketch dimensions necessary. *Autodesk Inventor* remembers which dimensions were added by the user and which were calculated by the system, so that automatic dimensions do not replace the desired dimensions.

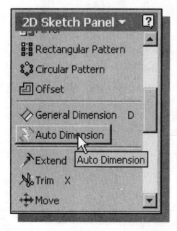

1. Click on the **Auto Dimension** icon in the *2D Sketch Panel*.

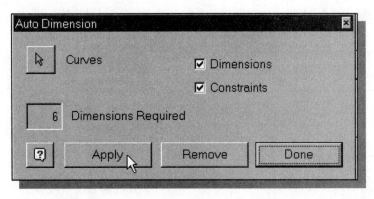

2. The *Auto Dimension* dialog box appears on the screen. Confirm that the **Dimensions** and **Constraints** options are switched *ON* as shown.

3. Click on the **Apply** button to proceed with the Auto Dimension command.

➢ Note that the system automatically calculates additional dimensions for all created geometric entities. Also note that **two** additional dimensions are still required to fully constrain the sketches as shown in the dialog box. The two missing dimensions are the location dimensions for the created sketches, to which we will apply proper constraints in the following section.

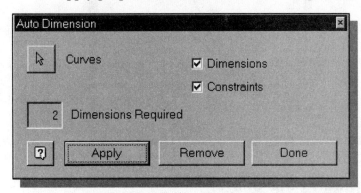

4. Click on the **Done** button to accept the created dimensions.

5. Reposition the displayed dimensions, by click and drag with the left-mouse-button, so that the dimensions can be more easily identified.

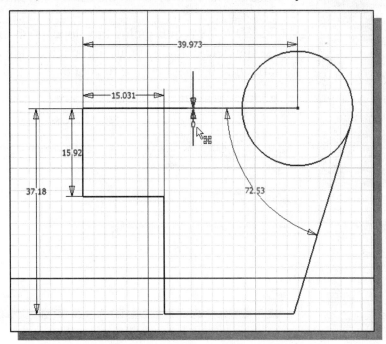

6. On your own, delete the angle dimension, trim the circle and add the length dimension as shown.

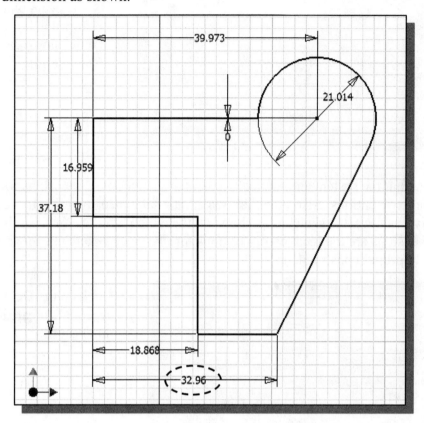

Creating Fillets and Completing the Sketch

1. Click on the **Fillet** icon in the *2D Sketch Panel*.

2. The *2D Fillet* radius dialog box appears on the screen. Use the default radius value and click on the top horizontal line and the arc to create a fillet as shown.

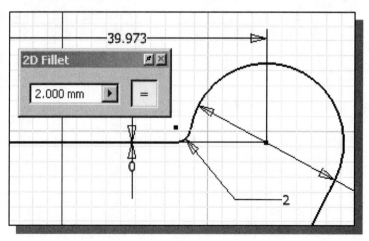

3. On your own, create the three additional fillets as shown in the below figure. Note that the **Equal** constraint is activated and all rounds and fillets are created with the constraint.

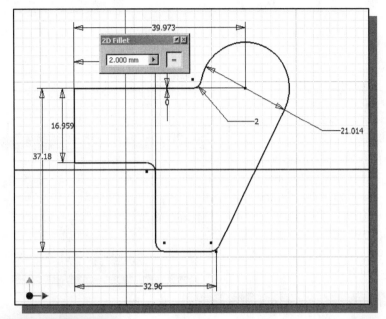

4. Click on the [**X**] icon to close the *2D Fillet* dialog box and end the **Fillet** command.

Fully Constrained Geometry

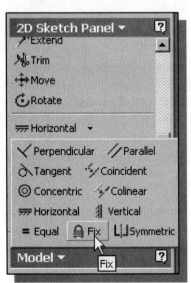

1. Click on the **Fix** constraint icon in the *2D Sketch Panel*.

2. Apply the **Fix** constraint to the center of the arc as shown below.

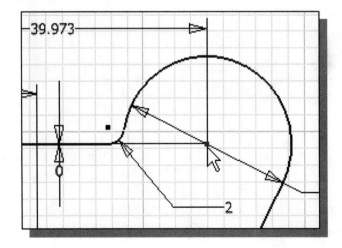

➤ Note that the sketch is fully constrained with the added constraint.

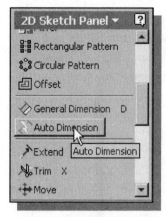

3. Click on the **Auto Dimension** icon in the *2D Sketch Panel*.

➢ Note that the system automatically recalculates if any additional dimension is needed for all created geometric entities. The applied Fix constraint provided the previously missing location dimensions for the created sketches.

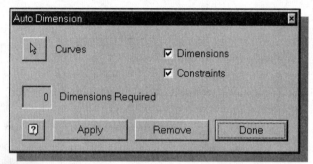

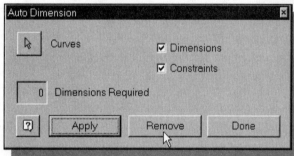

4. Click on the **Remove** button to proceed with the removal of the dimensions calculated by the system.

➢ Note that the system removes only those dimensions that were calculated and created by the system; all of the dimensions that we created are maintained correctly as shown.

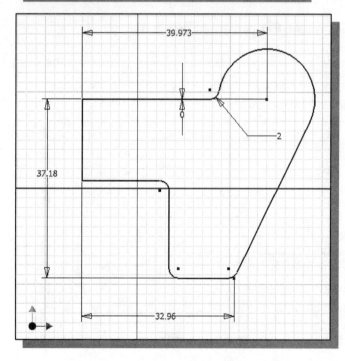

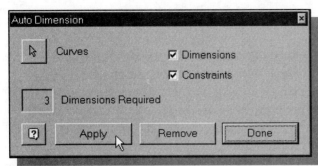

5. Click the **Apply** button to have the system recalculate and apply the missing dimensions.

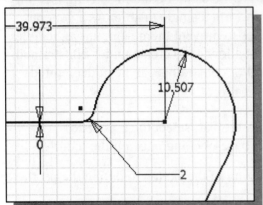

➢ Note that a radius dimension, instead of the diameter dimension that was previously placed, is displayed for the upper arc of the sketch.

6. Click **Done** to exit the Auto Dimension command.

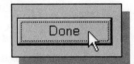

7. On your own, change the dimensions to the desired values as shown below. (Hint: Change the larger values first.)

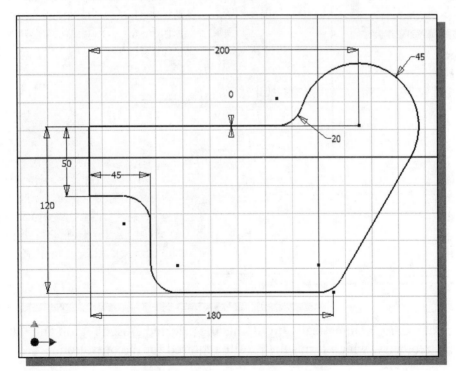

❖ Note that by applying proper geometric and dimensional constraints to the sketched geometry, a *constraint network* is created that assures the geometry shape behaves predictably as changes are made.

Profile Sketch

❖ In *Autodesk Inventor*, **profiles** are closed regions that are defined from sketches. Profiles are used as cross sections to create solid features. For example, **Extrude**, **Revolve**, **Sweep**, **Loft**, and **Coil** operations all require the definition of at least a single profile. The sketches used to define a profile can contain additional geometry since the additional geometry entities are consumed when the feature is created. To create a profile we can create single or multiple closed regions, or we can select existing solid edges to form closed regions. A profile cannot contain self-intersecting geometry; regions selected in a single operation form a single profile. As a general rule, we should dimension and constrain profiles to prevent them from unpredictable size and shape changes. *Autodesk Inventor* does allow us to create under-constrained or non-constrained profiles; the dimensions and/or constraints can be added/edited later.

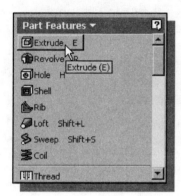

1. In the *Part Features* toolbar, select the **Extrude** command by left-mouse-clicking once on the icon.

2. Select the inside region of the sketch to define the profile of the extrusion.

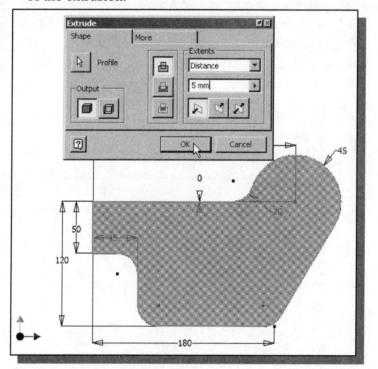

❖ *Autodesk Inventor* automatically highlights the selected closed region and the defining geometry, which forms the profile required for the operation.

3. In the *Extrude* dialog box, enter **5 mm** as the extrusion distance as shown.

4. Click on the **OK** button to accept the settings and create the solid feature.

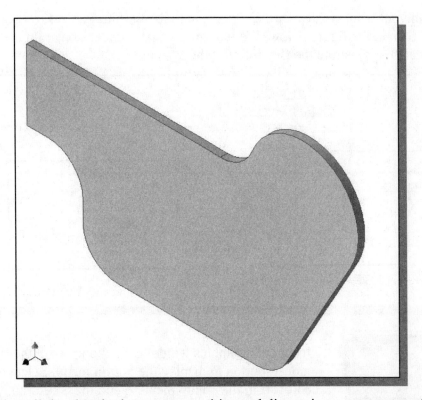

❖ Note that all the sketched geometry entities and dimensions are consumed and have disappeared from the screen when the feature is created.

Redefining the Sketch and Profile

- Engineering designs usually go through many revisions and changes. *Autodesk Inventor* provides an assortment of tools to handle design changes quickly and effectively. We will demonstrate some of the tools available by changing the base feature of the design. The profile used to create the extrusion is selected from the sketched geometry entities. In *Autodesk Inventor*, any profile can be edited and/or redefined at any time. It is this type of functionality in parametric solid modeling software that provides designers with greater flexibility and the ease to experiment with different design considerations.

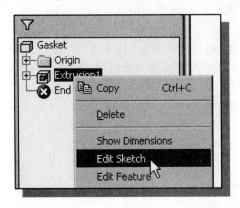

1. In the *Browser*, **right-mouse-click** once on the *Extruded* feature to bring up the option menu, then pick **Edit Sketch** in the pop-up menu.

2. On your own, create a circle and a rectangle of arbitrary sizes, positioned as shown in the figure below. We will intentionally under-constrain the new sketch to illustrate the flexibility of the system.

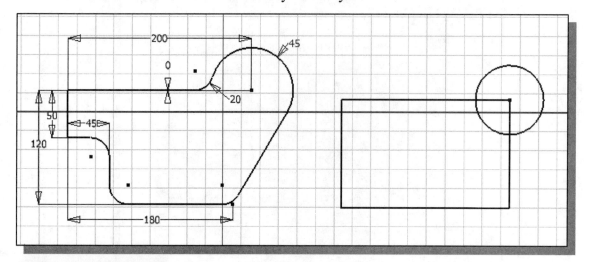

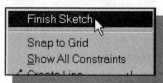

3. Inside the graphics window, click once with the right-mouse-button to display the option menu. Select **Finish Sketch** in the popup menu to end the Sketch option.

➢ Note that, at this point, the solid model remains the same as before. We will next redefine the profile used for the base feature.

4. In the *Browser*, right-mouse-click once on the *Extrusion1* feature to bring up the option menu, and then pick **Edit Feature** in the pop-up menu.

❖ *Autodesk Inventor* will now display the 2D sketch of the selected feature in the graphics window. We have literally gone back in time to the point where we define the extrusion feature.

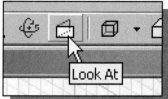

5. Click on the **Look At** icon in the *Standard* toolbar area.

• The **Look At** command automatically aligns the *sketch plane* of a selected entity to the screen.

6. Select any line segment of the 2D sketch.

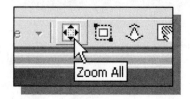

7. Click on the **Zoom All** icon in the *Standard* toolbar.

➤ We have literally gone back in time to the point where we first defined the 2D profile. The original sketch and the new sketch we just created are wireframe entities recorded by the system as belonging to the same **SKETCH**; but only the **PROFILED** entities are used to create the feature.

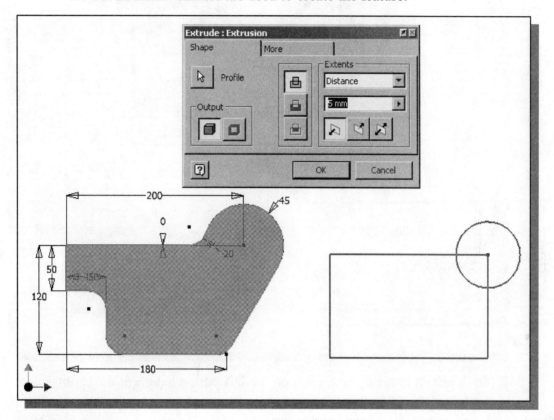

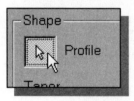

8. In the *Extrude* dialog box, click on the **Profile** button to edit/redefine the 2D profile.

9. The geometry entities used to define the profile are highlighted as shown in the figure above. To deselect, move the cursor inside the highlighted region, press the [**Ctrl**] key, and left-mouse-click once.

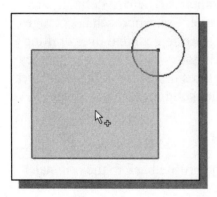

10. Click inside the rectangle of the sketch on the right side of the graphics window.

➤ *Autodesk Inventor* automatically selects the geometry entities that form a closed region to define the profile.

11. Click inside the circle to complete the profile definition as shown in the figure below.

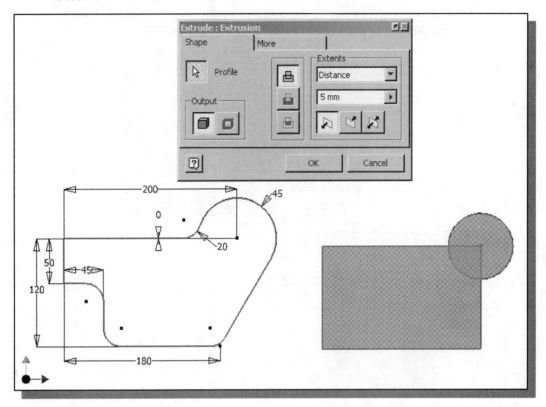

12. In the *Extrude* dialog box, click on the **OK** button to accept the settings and update the solid feature.

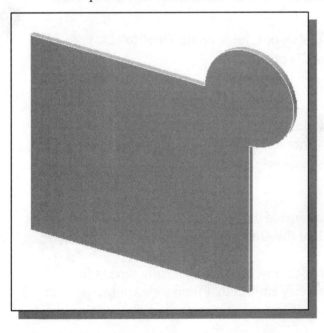

- The feature is recreated using the newly sketched geometric entities, which are under constrained and with no dimensions. The profile is created with extra wireframe entities by selecting multiple regions. The extra geometry entities can be used as construction geometry to help in defining the profile. This approach encourages engineering content over drafting technique, which is one of the key features of *Autodesk Inventor* over other solid modeling software.

13. On your own, repeat the above steps and set the profile back to the original sketch as shown on page 6-18.

Create an OFFSET Cut Feature

> ➢ To complete the design, we will create a cutout feature by using the **Offset** command. First we will set up the sketching plane to align with the front face of the 3D model.

1. In the *Standard* toolbar select the **Sketch** command by left-clicking once on the icon.

2. In the *Status Bar* area, the message: *"Select face, workplane, sketch or sketch geometry."* is displayed. Select the front face of the 3D model in the graphics window.

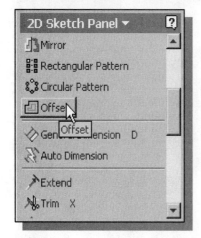

3. Click on the **Offset** icon in the *2D Sketch Panel*.

4. Select any edge of the front face of the 3D model. *Autodesk Inventor* will automatically select all of the connecting geometry to form a closed region.

5. Move the cursor toward the center of the selected region and notice an offset copy of the outline is displayed.

6. Left-mouse-click once to create the offset profile as shown.

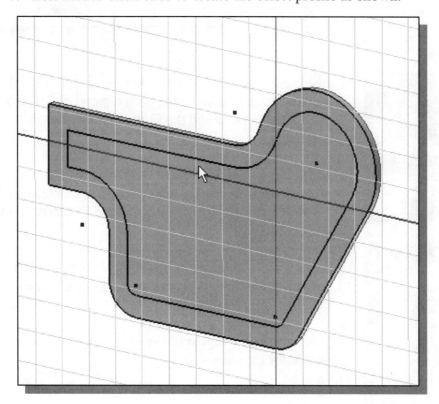

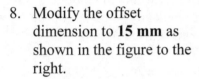

7. On your own, use the **General Dimension** command to create the offset dimension as shown in the figure below.

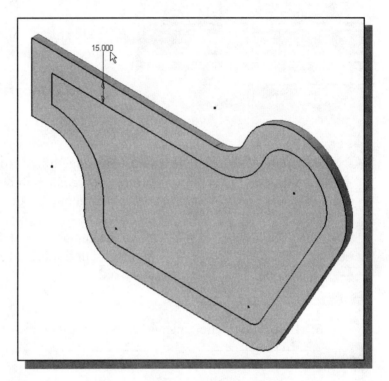

8. Modify the offset dimension to **15 mm** as shown in the figure to the right.

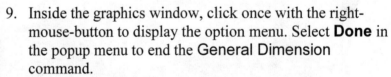

9. Inside the graphics window, click once with the right-mouse-button to display the option menu. Select **Done** in the popup menu to end the General Dimension command.

10. Inside the graphics window, click once with the right-mouse-button to display the option menu. Select **Finish Sketch** in the popup menu to end the Sketch option.

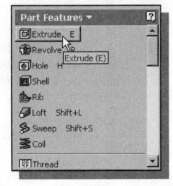

11. In the *Part Features* toolbar (the toolbar that is located to the left side of the graphics window), select the **Extrude** command by releasing the left-mouse-button on the icon.

12. Select the inside region of the offset geometry as the profile for the extrusion.

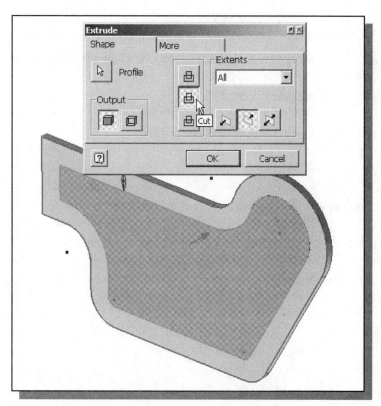

13. Inside the *Extrude* dialog box, select the **Cut** operation, set the *Extents* to **All** and set the direction of the cutout as shown in the figure to the left.

14. In the *Extrude* dialog box, click on the **OK** button to accept the settings and create the solid feature.

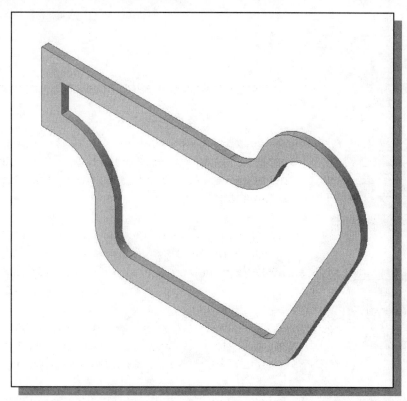

➤ The offset geometry is associated with the original geometry. On your own, adjust the overall height of the design to **150 millimeters** and confirm that the offset geometry is adjusted accordingly.

Questions:

1. What are the two types of wireframe geometry available in *Autodesk Inventor*?

2. Can we create a profile with extra 2D geometry entities?

3. How do we access the *Autodesk Inventor's* **Edit Sketch** option?

4. How do we create a *profile* in *Autodesk Inventor*?

5. Can we build a profile that consists of self-intersecting curves?

6. Describe the procedure to create a copy of a sketched 2D wireframe geometry?

7. Identify and briefly describe the following commands:

(a)

(b)

(c)

(d)

Exercises:

1. Dimensions are in inches. Plate Thickness: 0.25

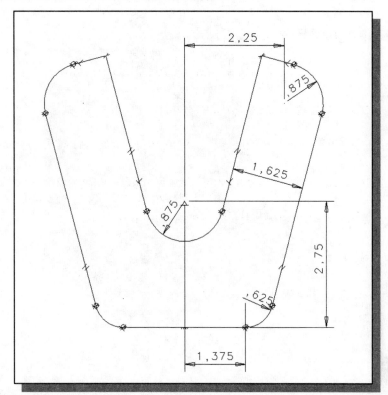

2. Dimensions are in millimeters.

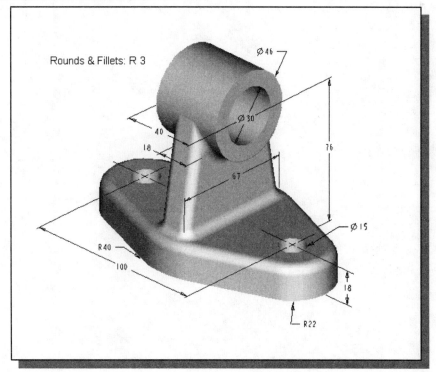

3. Dimensions are in inches.

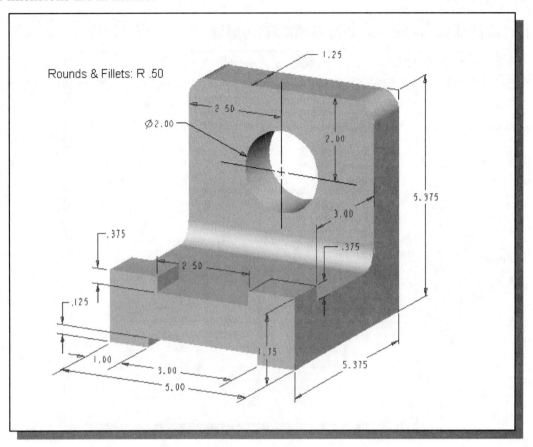

Chapter 7
Parent/Child Relationships and the BORN Technique

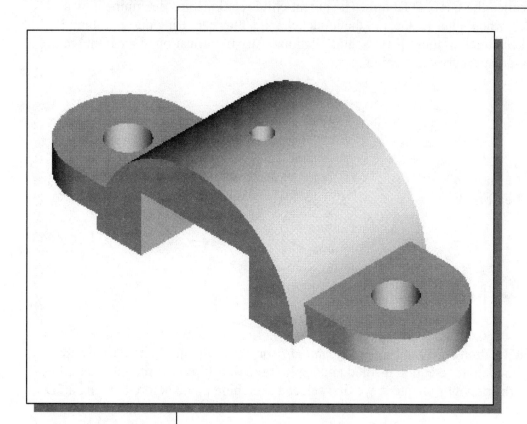

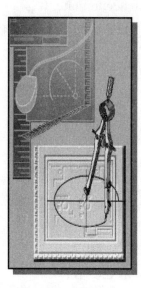

Learning Objectives

- ♦ **Understand the Concept and Usage of the BORN Technique**
- ♦ **Understand the Importance of Parent/Child Relations in Features**
- ♦ **Use the Suppress Feature Option**
- ♦ **Resolve Undesired Feature Interactions**

Introduction

The parent/child relationship is one of the most powerful aspects of *parametric modeling*. In *Autodesk Inventor*, each time a new modeling event is created, previously defined features can be used to define information such as size, location, and orientation. The referenced features become **PARENT** features to the new feature, and the new feature is called the **CHILD** feature. The parent/child relationships determine how a model reacts when other features in the model change, thus capturing design intent. It is crucial to keep track of these parent/child relations. Any modification to a parent feature can change one or more of its children.

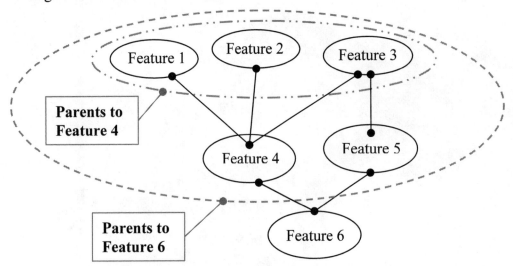

Parent/child relationships can be created *implicitly* or *explicitly*; implicit relationships are implied by the feature creation method and explicit relationships are entered manually by the user. In the previous chapters, we first select a sketching plane before creating a 2D profile. The selected surface becomes a parent of the new feature. If the sketching plane is moved, the child feature will move with it. As one might expect, parent/child relationships can become quite complicated when the features begin to accumulate. It is therefore important to think about modeling strategy before we start to create anything. The main consideration is to try to plan ahead for possible design changes that might occur which would be affected by the existing parent/child relationships. Parametric modeling software, such as Autodesk Inventor, also allows us to adjust feature properties so that any feature conflicts can be quickly resolved.

The BORN Technique

In the previous chapters, we have chosen the *first feature* to be an extruded solid object. All subsequent features, therefore, are built by referencing the first feature, or *base feature*. The *base feature* is the center of all features and is considered as the key feature of the design. This approach places much emphasis on the selection of the *base feature*. In most cases, this approach is quite adequate and proper in creating the solid models.

A more advanced technique of creating solid models is what is known as the "**Base Orphan Reference Node**" (BORN) technique. The basic concept of the BORN technique is to use a *Cartesian coordinate system* as the first feature prior to creating any solid features. With the *Cartesian coordinate system* established, we then have three mutually perpendicular datum planes (namely the *XY, YZ*, and *ZX planes*) available to use as sketching planes. The three datum planes can also be used as references for dimensions and geometric constructions. Using this technique, the first node in the history tree is called an "orphan," meaning that it has no history to be replayed. The technique of creating the reference geometry in this "base node" is therefore called the "Base Orphan Reference Node" (BORN) technique.

Autodesk Inventor automatically establishes a set of reference geometry when we start a new part, namely a *Cartesian coordinate system* with three work planes, three work axes, and a work point All subsequent solid features can then use the coordinate system and/or reference geometry as sketching planes. The *base feature* is still important, but the *base feature* is no longer the ONLY choice for selecting the sketching plane for subsequent solid features. This approach provides us with more options while we are creating parametric solid models. More importantly, this approach provides greater flexibility for part modifications and design changes. This approach is also very useful in creating assembly models, which will be illustrated in *Chapter 12*.

The *U-Bracket* Design

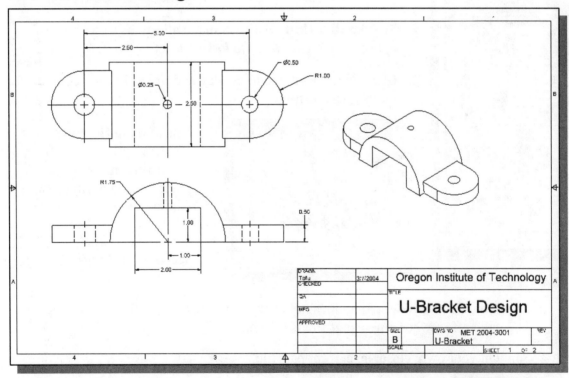

> Based on your knowledge of *Inventor* so far, how many features would you use to create the model? Which feature would you choose as the **base feature**? What is your choice for arranging the order of the features? Would you organize the features differently if the rectangular cut at the center is changed to a circular shape (Radius: 80mm)?

Starting Autodesk Inventor

1. Select the *Autodesk Inventor* option on the *Start* menu or select the *Autodesk Inventor* icon on the desktop to start *Autodesk Inventor*. The *Autodesk Inventor* main window will appear on the screen.

2. Once the program is loaded into the memory, the ***Startup*** dialog box appears at the center of the screen.

3. Select the **New** icon with a single click of the left-mouse-button in the *What to Do* dialog box.

4. Select the **English** tab and in the *New - Choose Template* area, select **Standard(in).ipt**.

5. Pick **OK** in the *Startup* dialog box to accept the selected settings.

Default Sketch Plane Setting

1. Select **Application Options** in the **Tools** pull-down menu.

• The **Application Options** menu allows us to set behavioral options, such as *color*, *file locations*, etc.

2. Click on the **Part** tab to display and/or modify the default sketch plane settings.

3. Click on the **No new sketch** option to establish the setting of which sketch plane is used during new part creation.

4. Click on the **OK** button to accept the setting.

• Note that the new setting does not change the current part file; the setting will take effect when a new part file is opened.

5. On your own, **close** the current part file.

6. Click on the **New** icon in the *Standard* toolbar and open a new *English* units standard part file.

❖ Note the new part file, *PART2*, is no longer set up with a default sketching plane.

Applying the BORN Technique

1. In the *Part Browser* window, click on the [**+**] symbol in front of the ***Origin*** feature to display more information on the feature.

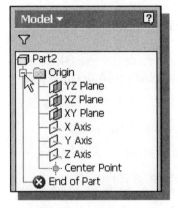

❖ In the *Part Browser* window, notice a new part name appeared with seven work features established. The seven work features include three *workplanes*, three *work axes*, and a *work point*. By default, the three workplanes and work axes are aligned to the **world coordinate system** and the work point is aligned to the *origin* of the **world coordinate system**.

2. Inside the *Browser* window, move the cursor on top of the third work plane, the ***XY Plane***. Notice a rectangle, representing the work-plane, appears in the graphics window.

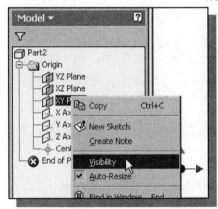

3. Inside the *Browser* window, click once with the right-mouse-button on *XY Plane* to display the option menu. Click on **Visibility** to toggle on the display of the plane.

4. On your own, repeat the above steps and toggle ***ON*** the display of all of the *workplanes*, *work axes*, and the *center point* on the screen.

5. On your own, use the *Dynamic Viewing* options (3D Rotate, Zoom and Pan) to view the work features established.

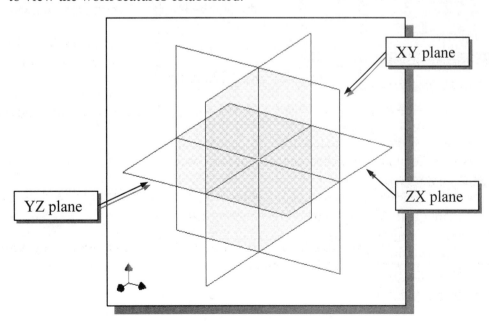

❖ By default, the basic set of workplanes is aligned to the world coordinate system; the workplanes are the first features of the part. We can now proceed to create solid features referencing the three mutually perpendicular datum planes. Instead of using only the default sketching plane as the starting point, we can now select any of the work planes as the sketching planes for subsequent solid features.

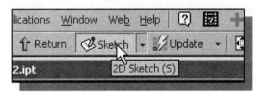

6. In the *Standard* toolbar select the **Sketch** command by left-clicking once on the icon.

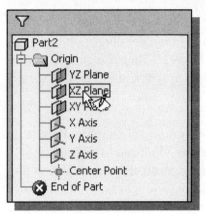

7. In the *Status Bar* area, the message: "*Select face, workplane, sketch or sketch geometry.*" is displayed. *Autodesk Inventor* expects us to identify a planar surface where the 2D sketch of the next feature is to be created. Move the graphics cursor on top of *XZ Plane*, inside the *Browser* window as shown, and notice that *Autodesk Inventor* will automatically highlight the corresponding plane in the graphics window. Left-click once to select the *XZ Plane* as the sketching plane.

❖ *Autodesk Inventor* allows us to identify and select features in the graphics window as well as in the *Browser* window.

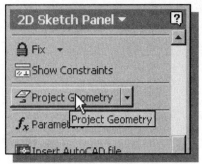

8. Select the **Project Geometry** command in the *2D Sketch Panel*. The **Project Geometry** command allows us to project existing features to the active sketching plane. Left-click once on the icon to activate the **Project Geometry** command.

9. In the *Status Bar* area, the message: "*Select edge, vertex, work geometry or sketch geometry to project.*" is displayed. *Autodesk Inventor* expects us to select any existing geometry, which will be projected onto the active sketching plane. Inside the *Browser* window, move the graphics cursor on top of *Center Point* as shown, and notice that *Autodesk Inventor* will automatically highlight the corresponding feature in the graphics window. Left-click once to project the center point onto the sketching plane.

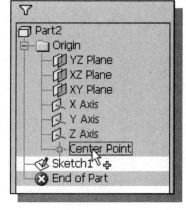

Creating the 2D Sketch for the Base Feature

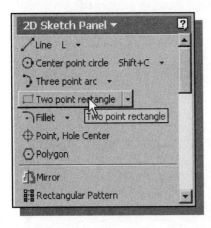

1. Select the **Two point rectangle** command by clicking once with the **left-mouse-button** on the icon in the *Sketch* toolbar.

2. Create a rectangle surrounding the projected center point as shown below.

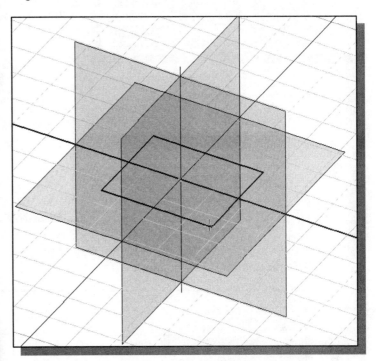

3. Select the **Center point circle** command by clicking once with the left-mouse-button on the icon in the *Sketch* toolbar.

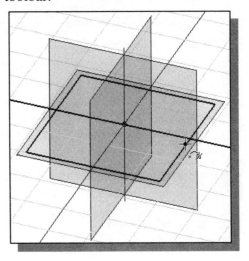

➤ We will align the center of the circle to the midpoint of the short edge of the rectangle.

4. Move the cursor along the shorter edge of the rectangle and pick the midpoint of the edge when the midpoint is displayed with a GREEN color.

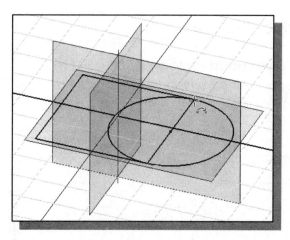

5. Select a neighboring corner of the rectangle to create a circle as shown below.

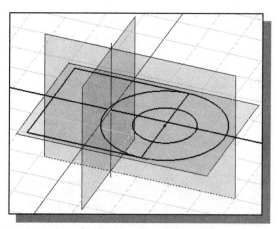

6. On your own, create a smaller circle sharing the same center point location as shown.

7. On your own, repeat the above steps and create two additional circles on the other side of the rectangle.

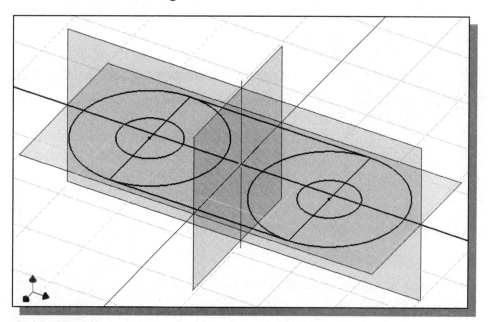

8. Inside the graphics window, click once with the right-mouse-button to display the option menu. Select **Done** in the popup menu to end the Circle command.

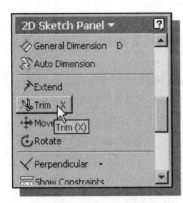

9. Choose **Trim** in the *2D Sketch Panel*. The message *"Trim curves"* is displayed in the prompt area.

10. On your own, trim the rectangle and the two outer circles so that the sketch appears as shown.

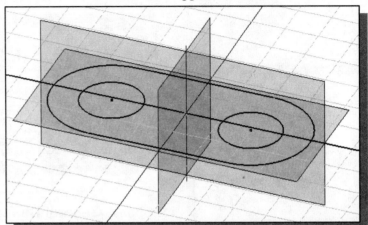

11. Move the cursor on top of the **General Dimension** icon. The General Dimension command allows us to quickly create and modify dimensions. Left-click once on the icon to activate the General Dimension command.

12. On your own, create the dimensions to fully constrain the sketch as shown. (Do not be overly concerned with the actual numbers displayed; the dimensions will be adjusted shortly.)

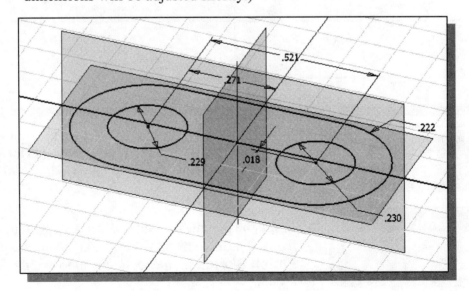

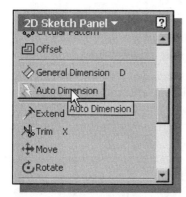

13. Click on the **Auto Dimension** icon in the *2D Sketch Panel*.

14. Confirm the 2D sketch is fully constrained as shown in the figure.

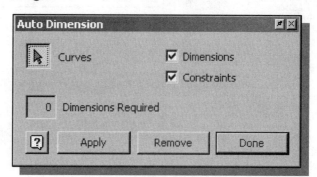

15. Choose **General Dimension** in the *2D Sketch Panel*.

16. On your own, adjust the dimensions as shown in the figure.

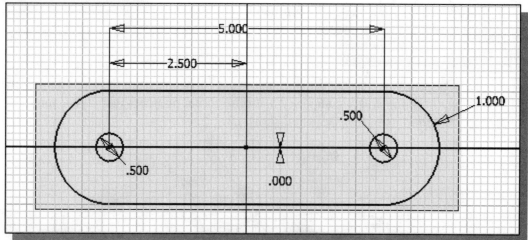

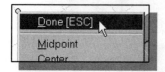

17. Inside the graphics window, click once with the right-mouse-button to display the option menu. Select **Done** in the popup menu to end the General Dimension command.

18. Inside the graphics window, click once with the right-mouse-button to display the option menu. Select **Finish Sketch** in the popup menu to end the Sketch option.

Create the First Extrude Feature

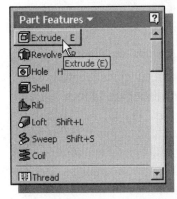

1. In the *Part Features* toolbar (the toolbar that is located to the left side of the graphics window), select the **Extrude** command by clicking once with the left-mouse-button on the icon.

2. Select the inside region of the sketch to define the profile of the extrusion as shown.

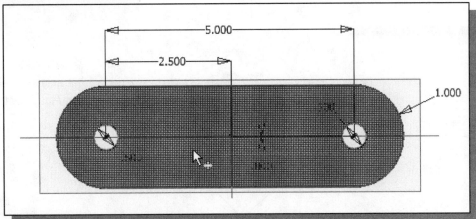

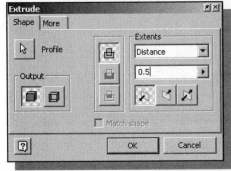

3. In the *Distance* option box, enter **0.5** as the extrusion distance.

4. In the *Extrude* popup window, click on the **OK** button to create the base feature.

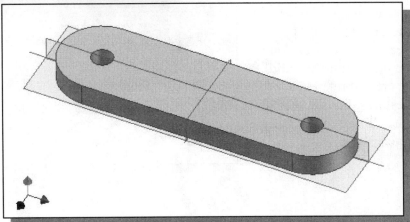

The Implied Parent/Child Relationships

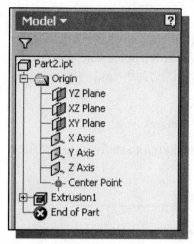

- The *Model Tree* shows two main features: the **Origin** (default work features) and the base feature (**Extrusion1** shown in the figure) we just created. The parent/child relationships were established implicitly when we created the base feature: (1) *XZ workplane* was selected as the sketch plane; (2) *Center Point* was used as the reference point to align the 2D sketch of the base feature.

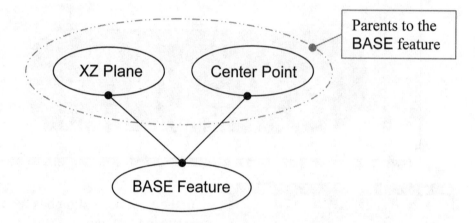

Creating the Second Solid Feature

- For the next solid feature, we will create the top section of the design. Note that the center of the base feature is aligned to the *Center Point* of the default work features. This was done intentionally so that additional solid features can be created referencing the default work features. For the second solid feature, the *XY workplane* will be used as the sketch plane.

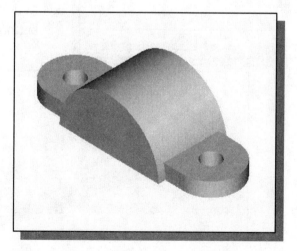

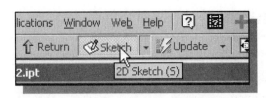

1. In the *Standard* toolbar select the **Sketch** command by left-clicking once on the icon.

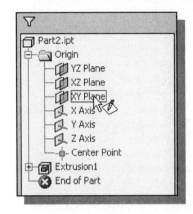

2. In the *Status Bar* area, the message: "*Select face, work plane, sketch or sketch geometry.*" is displayed. Pick the *XY plane* by clicking the work plane name inside the *Browser* as shown.

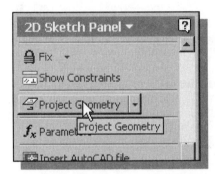

3. Select the **Project Geometry** command in the *2D Sketch Panel*. The Project Geometry command allows us to project existing features to the active sketching plane. Left-click once on the icon to activate the Project Geometry command.

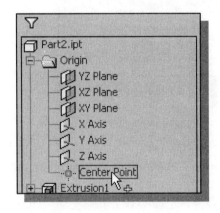

4. In the *Status Bar* area, the message: "*Select edge, vertex, work geometry or sketch geometry to project*" is displayed. Left-click once on *Center Point* to project the center point onto the sketching plane.

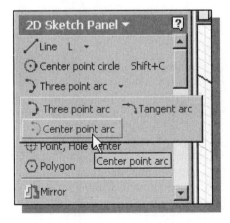

5. Select the **Center point arc** command by clicking once with the left-mouse-button on the icon in the *icon stack* as shown.

6. Pick the projected center point as the center location of the new arc.

7. On your own, create a semi-circle of arbitrary size, with both endpoints aligned to the X-axis, as shown below.

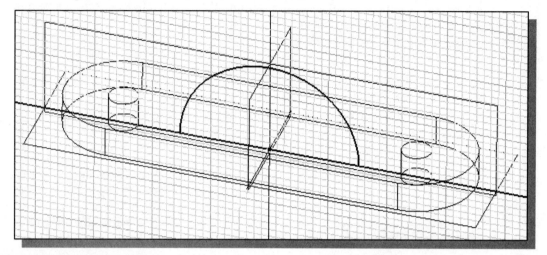

8. Move the cursor on top of the **General Dimension** icon. Left-click once on the icon to activate the General Dimension command.

9. On your own, create and adjust the radius of the arc to **1.75**.

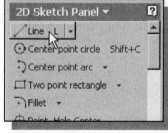

10. Select the **Line** command in the *2D Sketch* toolbar.

11. Create a line connecting the two endpoints of the arc as shown in the figure below.

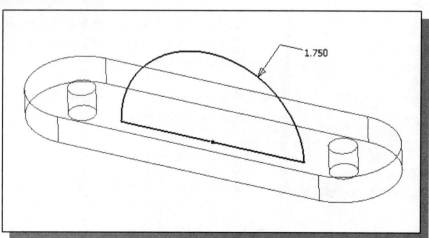

12. Inside the graphics window, click once with the right-mouse-button to display the option menu. Select **Finish Sketch** in the popup menu to end the Sketch option.

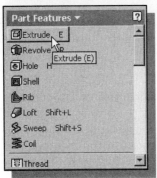

13. In the *Part Features* toolbar, select the **Extrude** command by clicking once with the left-mouse-button on the icon.

14. Select the inside region of the sketched arc-line curves as the profile to be extruded.

15. In the *Extrude* dialog box, set to the **Mid-Plane** option.

16. In the *Distance* value box, set the extrusion *distance* to **2.5**.

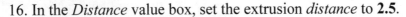
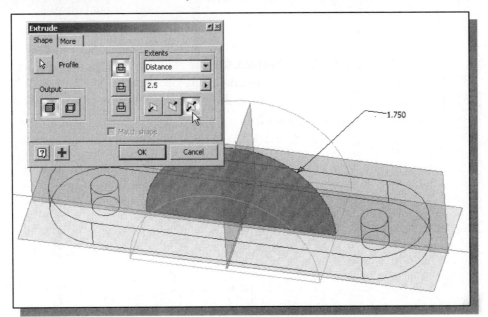

17. Click on the **OK** button to proceed with the **Extrude** operation.

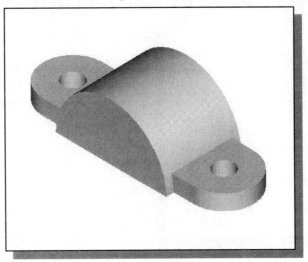

Creating the First Cut feature

- A rectangular cut will be created as the next solid feature.

1. In the *Standard* toolbar select the **Sketch** command by left-clicking once on the icon.

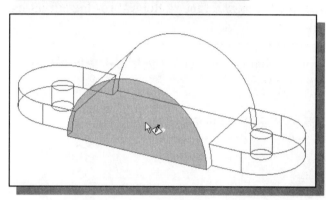

2. In the *Status Bar* area, the message: "*Select face, workplane, sketch or sketch geometry*" is displayed. Pick the front vertical face of the solid model shown.

3. On your own, create a rectangle and apply the dimensions as shown below. (Note that the **Center Point** of the default work features is projected to assure the alignment of the 2D sketch.)

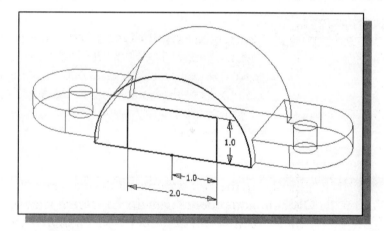

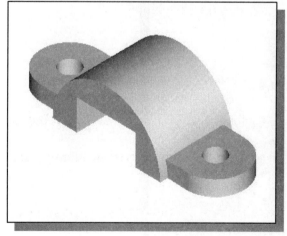

4. On your own, use the **Extrude** command and create a cutout that cuts through the entire 3D solid model as shown.

The Second Cut Feature

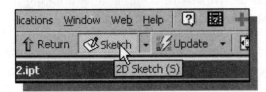

1. In the *Standard* toolbar select the **Sketch** command by left-clicking once on the icon.

2. In the *Status Bar* area, the message: "*Select face, work plane, sketch or sketch geometry*" is displayed. Select the horizontal face of the last cut feature as the *sketching plane*.

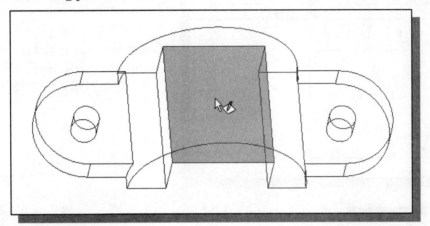

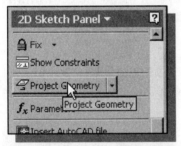

3. Select the **Project Geometry** command in the *2D Sketch Panel*. The Project Geometry command allows us to project existing features onto the active sketching plane. Left-click once on the icon to activate the **Project Geometry** command.

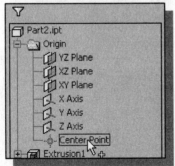

4. In the *Status Bar* area, the message: "*Select edge, vertex, work geometry or sketch geometry to project*" is displayed. Left-click once on **Center Point** to project the center point onto the sketching plane.

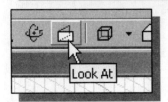

5. In the *Standard* toolbar area, click on the **Look At** button.

6. Select one of the edges of the highlighted sketching plane.

❖ The **Look At** command can be used to switch to a 2D view of a selected surface.

7. Select the **Center point circle** command by clicking once with the left-mouse-button on the icon in the *2D Sketch Panel*.

8. Pick the projected center point to align the center of the new circle.

9. Create a circle of arbitrary size.

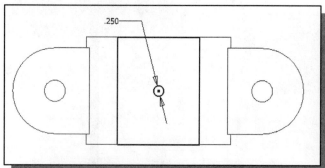

10. On your own, add the size dimension of the circle and set the dimension to **0.25**.

11. Inside the graphics window, click once with the right-mouse-button to display the option menu. Select **Finish Sketch** in the popup menu to end the Sketch option.

12. In the *Part Features* toolbar, select the **Extrude** command by clicking once with the left-mouse-button on the icon.

13. Select the inside region of the sketched circle as the profile to be extruded.

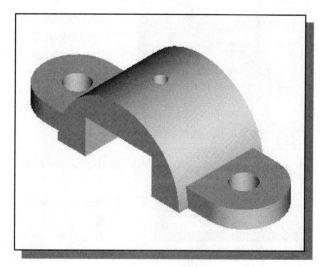

14. In the *Extrude* dialog box, set to the **Cut All** option.

15. Click on the **Done** button to proceed with creating the cut feature model.

Examining the Parent/Child Relationships

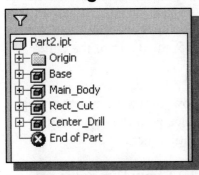

1. On your own, rename the feature names to: **Base**, **MainBody**, **Rect_Cut** and **Center_Drill** as shown in the figure.

❖ The *Model Tree* window now contains seven items: the **Origin** (default work features) and four solid features. All of the parent/child relationships were established implicitly as we created the solid features. As more features are created, it becomes much more difficult to make a sketch showing all the parent/child relationships involved in the model. On the other hand, it is not really necessary to have a detailed picture showing all the relationships among the features. In using a feature-based modeler, the main emphasis is to consider the interactions that exist between the **immediate features**. Treat each feature as a unit by itself, and be clear on the parent/child relationships for each feature. Thinking in terms of *features* is what distinguishes *feature-based modeling* and the previous generation solid modeling techniques. Let us take a look at the last feature we created, the **Center_Drill** feature. What are the parent/child relationships associated with this feature? (1) Since this is the last feature we created, it is not a parent feature to any other features. (2) Since we used one of the surfaces of the rectangular cutout as the sketching plane, the **Rect_Cut** feature is a parent feature to the **Center_Drill** feature. (3) We also used the projected Origin as a reference point to align the center; therefore, the **Origin** is also a parent to the **Center_Drill** feature.

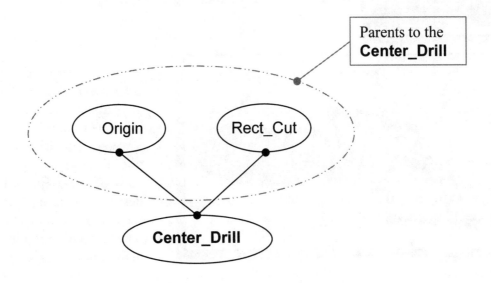

Modify a Parent Dimension

❖ Any changes to the parent features will affect the child feature. For example, if we modify the height of the **Rect_Cut** feature from 1.0 to 0.75, the depth of the child feature (**Center_Drill** feature) will be affected.

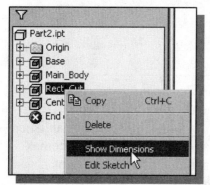

1. In the *Model Tree* window, move the cursor on top of **Rect_Cut** and click once with the right-mouse-button to bring up the option menu.

2. In the option menu, select **Show Dimensions.**

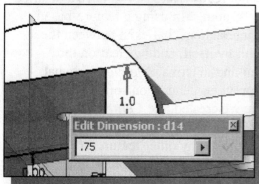

3. Select the height dimension (1.0) by double-clicking on the dimension.

4. Enter **0.75** as the new height dimension as shown.

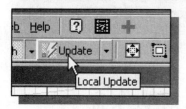

5. Click on the **Update** button in the *Standard* toolbar area to proceed with updating the solid model.

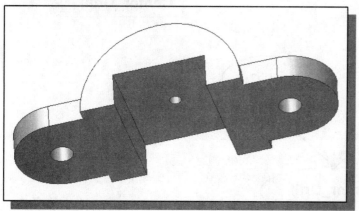

➢ Note that the position of the **Center_Drill** feature is also adjusted as the placement plane is lowered. The drill-hole still goes through the main body of the *U-Bracket* design. The parent/child relationship assures the intent of the design is maintained.

6. On your own, adjust the height of the **Rect_Cut** feature back to **1.0** inch before proceeding to the next section.

A Design Change

➢ Engineering designs usually go through many revisions and changes. For example, a design change may call for a circular cutout instead of the current rectangular cutout feature in our model. *Autodesk Inventor* provides an assortment of tools to handle design changes quickly and effectively. In the following sections, we will demonstrate some of the more advanced tools available in *Autodesk Inventor*, which allow us to perform the modification of changing the rectangular cutout (2.0 × 1.0 inch) to a circular cutout (radius: 1.25 inch).

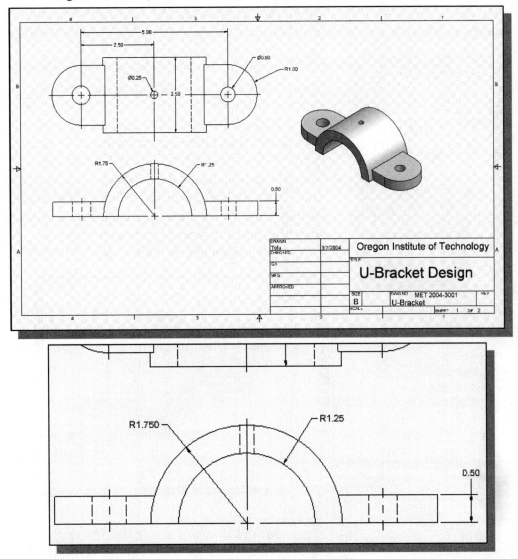

❖ Based on your knowledge of *Autodesk Inventor* so far, how would you accomplish this modification? What other approaches can you think of that are also feasible? Of the approaches you came up with, which one is the easiest to do and which is the most flexible? If this design change were anticipated right at the beginning of the design process, what would be your choice in arranging the order of the features? You are encouraged to perform the modifications prior to following through the rest of the tutorial.

Feature Suppression

❖ With *Autodesk Inventor*, we can take several different approaches to accomplish this modification. We could (1) create a new model, or (2) change the shape of the existing cut feature using the **Redefine** command, or (3) perform **feature suppression** on the rectangular cut feature and add a circular cut feature. The third approach offers the most flexibility and requires the least amount of editing to the existing geometry. **Feature suppression** is a method that enables us to disable a feature while retaining the complete feature information; the feature can be reactivated at any time. Prior to adding the new cut feature, we will first suppress the rectangular cut feature.

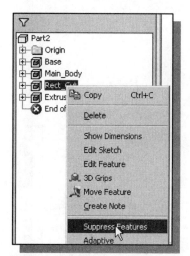

1. Move the cursor inside the *Model Tree* window. Click once with the right-mouse-button on top of **Rect_Cut** to bring up the option menu.

2. Pick **Suppress** in the pop-up menu.

❖ We have literally *gone back in time*. The *Rect_Cut* and Center_Drill features have disappeared in the display area. The child feature cannot exist without its parent(s), and any modification to the parent (*Rect_Cut*) influences the child *(Center_Drill)*.

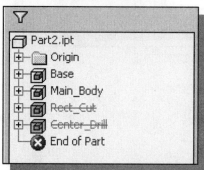

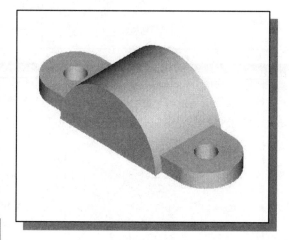

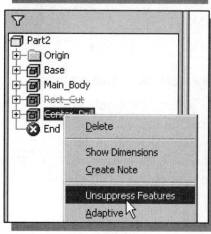

3. Move the cursor inside the *Model Tree* window. Click once with the right-mouse-button on top of **Center_Drill** to bring up the option menu.

4. Pick **Unsuppress Features** in the pop-up menu.

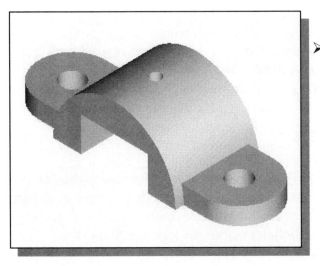

> In the display area and the *Model Tree* window, both the **Rect_Cut** feature and the **Center_Drill** feature are re-activated. The child feature cannot exist without its parent(s); the parent *(Rect_Cut)* must be activated to enable the child *(Center_Drill)*.

A Different Approach to the CENTER_DRILL Feature

❖ The main advantage of using the BORN technique is to provide greater flexibility for part modifications and design changes. In this case, the *Center_Drill* feature can be placed on the *XZ workplane* and therefore not be linked to the *Rect_Cut* feature.

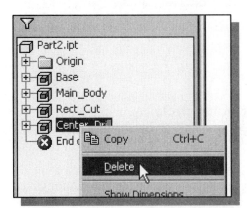

1. Move the cursor inside the *Model Tree* window. Click once with the right-mouse-button on top of *Center_Drill* to bring up the option menu.

2. Pick **Delete** in the pop-up menu.

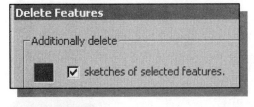

3. In the *Delete Features* window, confirm the **sketches of selected features** option is switched *ON*.

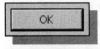

4. Click **OK** to proceed with the Delete command.

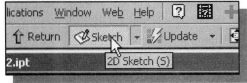

5. In the *Standard* toolbar select the **Sketch** command by left-clicking once on the icon.

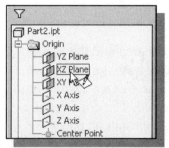

6. In the *Status Bar* area, the message: "*Select face, workplane, sketch or sketch geometry*" is displayed. Pick the **XZ workplane** in the *Model Tree* window as shown below.

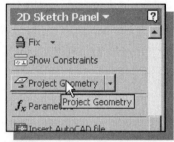

7. Select the **Project Geometry** command in the *2D Sketch Panel*. The Project Geometry command allows us to project existing features onto the active sketching plane. Left-click once on the icon to activate the Project Geometry command.

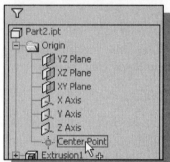

8. In the *Status Bar* area, the message: "*Select edge, vertex, work geometry or sketch geometry to project*" is displayed. Left-click once on **Center Point** to project the center point onto the sketching plane.

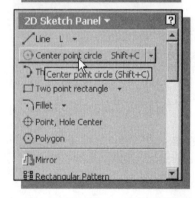

9. Select the **Center point circle** command by clicking once with the left-mouse-button on the icon in the *2D Sketch Panel*.

10. Pick the projected center point to align the center of the new circle.

11. Create a circle of arbitrary size.

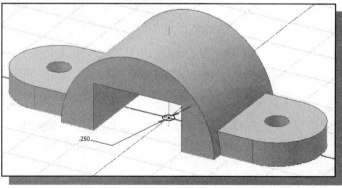

12. On your own, add the size dimension of the circle and set the dimension to **0.25**.

13. On your own, complete the extrude cut feature.

Suppress the Rect_Cut Feature

❖ Now the *Center_Drill* feature is no longer a child of the *Rect_Cut* feature, any changes to the *Rect_Cut* feature does not affect the *Center_Drill* feature.

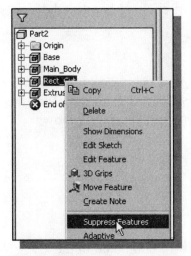

1. Move the cursor inside the *Model Tree* window. Click once with the right-mouse-button on top of *Rect_Cut* to bring up the option menu.

2. Pick **Suppress Features** in the pop-up menu.

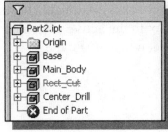

❖ The *Rect_Cut* feature is now disabled without affecting the *Center_Drill* feature.

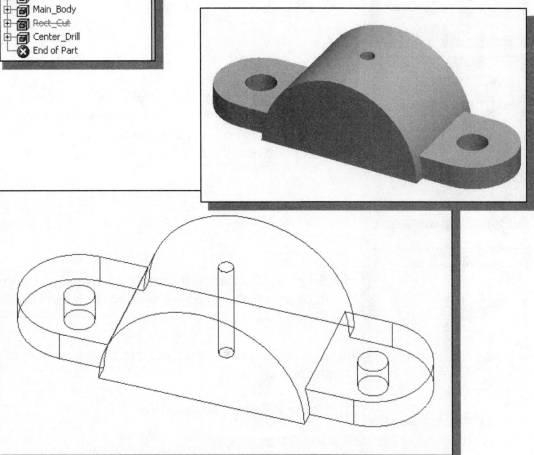

Creating a Circular Cut Feature

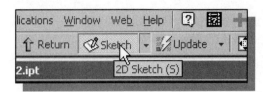

1. In the *Standard* toolbar select the **Sketch** command by left-clicking once on the icon.

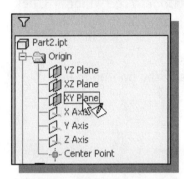

2. In the *Status Bar* area, the message: "*Select face, work plane, sketch or sketch geometry*" is displayed. Pick the *XY workplane* in the *Model Tree* window as shown below.

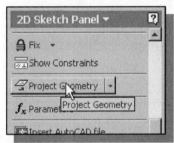

3. Select the **Project Geometry** command in the *2D Sketch Panel*. The Project Geometry command allows us to project existing features onto the active sketching plane. Left-click once on the icon to activate the Project Geometry command.

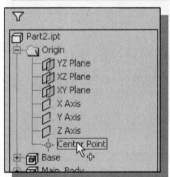

4. In the *Status Bar* area, the message: "*Select edge, vertex, work geometry or sketch geometry to project*" is displayed. Left-click once on *Center Point* to project the center point onto the sketching plane.

5. Select the **Center point circle** command by clicking once with the left-mouse-button on the icon in the *2D Sketch Panel*.

6. Pick the projected center point to align the center of the new circle.

7. Create a circle of arbitrary size.

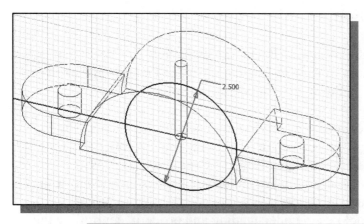

8. On your own, create the size dimension of the circle and set the dimension to **2.5** as shown in the figure.

9. On your own, complete the cut feature using the **Extrude** command.

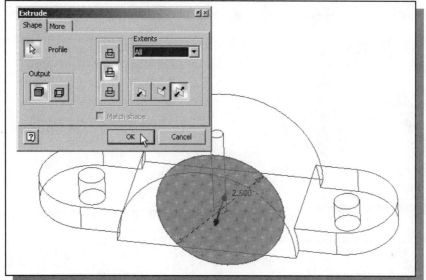

❖ Note that the parents of the *Circular_Cut* feature are the *XY Plane* and the projected *Center Point*.

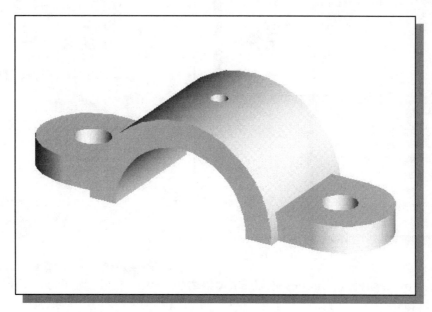

A Flexible Design Approach

In a typical design process, the initial design will undergo many analyses, testing, reviews and revisions. *Autodesk Inventor* allows the users to quickly make changes and explore different options of the initial design throughout the design process.

The model we constructed in this chapter contains two distinct design options. The *feature-based parametric modeling* approach enables us to quickly explore design alternatives and we can include different design ideas into the same model. With parametric modeling, designers can concentrate on improving the design and the design process to be much more quickly and effortlessly. The key to successfully using parametric modeling as a design tool lies in understanding and properly controlling the interactions of features, especially the parent/child relations.

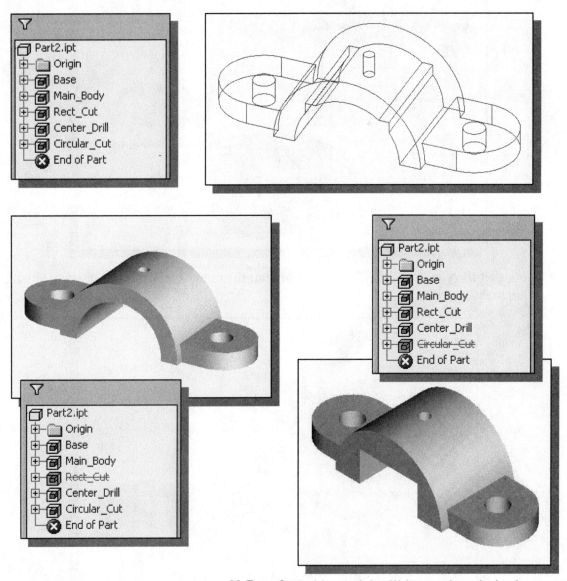

➤ On your own, save the model as ***U-Bracket***; this model will be used again in the next chapter.

Questions:

1. Why is it important to consider the parent/child relationships in between features?

2. Describe the procedure to **suppress** a *feature*.

3. What is the basic concept of the BORN technique?

4. What happen to a feature when it is suppressed?

5. How do you identify a suppressed feature in a model?

6. What is the main advantage of using the BORN technique?

7. Create sketches showing the steps you plan to use to create the models shown on the next page:

Exercises: (Dimensions are in inches)

1.

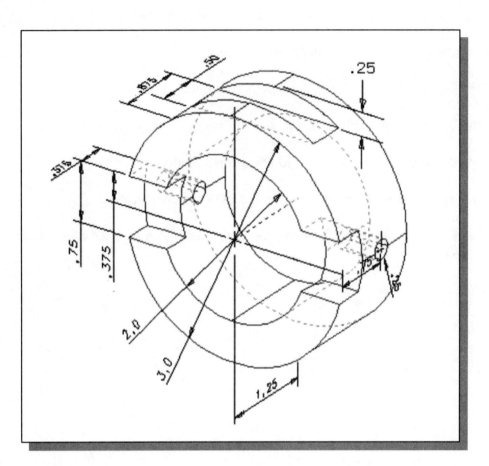

2.

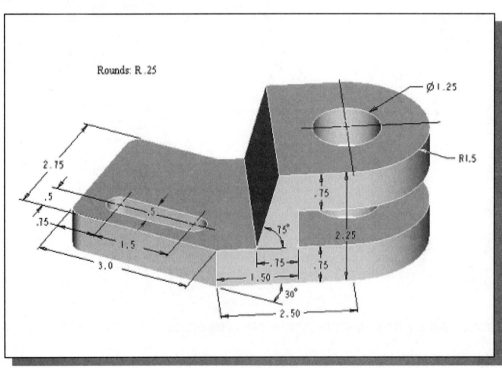

3.

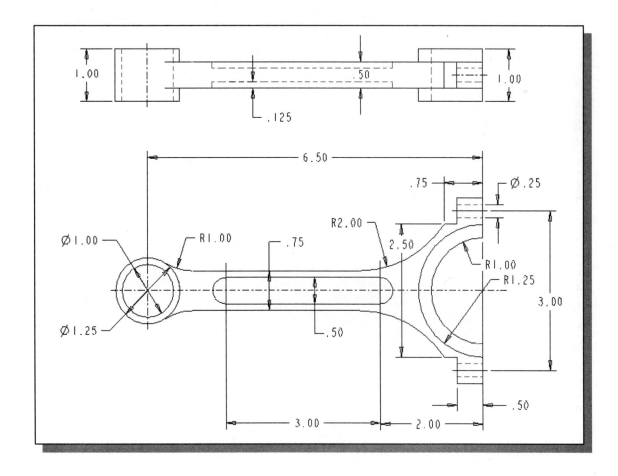

NOTES:

Chapter 8
Part Drawings and Associative Functionality

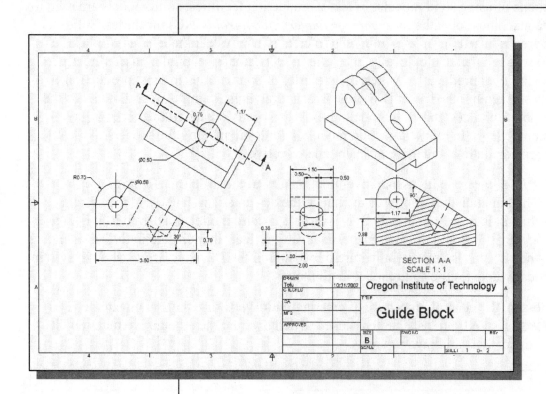

Learning Objectives

- ♦ **Create Drawing Layouts from Solid Models**
- ♦ **Understand Associative Functionality**
- ♦ **Using the default Borders and Title Block in the Layout Mode**
- ♦ **Arrange and Manage 2D Views in Drawing Mode**
- ♦ **Display and Hide Feature Dimensions**
- ♦ **Create Reference Dimensions**

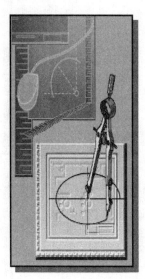

Drawings from Parts and Associative Functionality

With the software/hardware improvements in solid modeling, the importance of two-dimensional drawings is decreasing. Drafting is considered one of the downstream applications of using solid models. In many production facilities, solid models are used to generate machine tool paths for *computer numerical control* (CNC) machines. Solid models are also used in *rapid prototyping* to create 3D physical models out of plastic resins, powdered metal, etc. Ideally, the solid model database should be used directly to generate the final product. However, the majority of applications in most production facilities still require the use of two-dimensional drawings. Using the solid model as the starting point for a design, solid modeling tools can easily create all the necessary two-dimensional views. In this sense, solid modeling tools are making the process of creating two-dimensional drawings more efficient and effective.

Autodesk Inventor provides associative functionality in the different *Autodesk Inventor* modes. This functionality allows us to change the design at any level, and the system reflects it at all levels automatically. For example, a solid model can be modified in the *Part Modeling Mode* and the system automatically reflects that change in the *Drawing Mode*. And we can also modify a feature dimension in the *Drawing Mode*, and the system automatically updates the solid model in all modes.

In this lesson, the general procedure of creating multi-view drawings is illustrated. The *U_Bracket* design from last chapter is used to demonstrate the associative functionality between the model and drawing views.

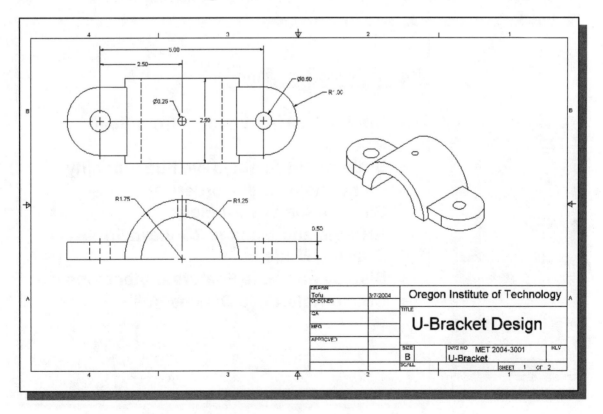

Starting Autodesk Inventor

1. Select the **Autodesk Inventor** option on the *Start* menu or select the **Autodesk Inventor** icon on the desktop to start *Autodesk Inventor*. The *Autodesk Inventor* main window will appear on the screen.

2. Once the program is loaded into memory, the *Startup* dialog box appears at the center of the screen.

3. Select **Open** with a single click of the left-mouse-button in the *What to Do* dialog box.

4. In the *File name* list box, select the **U-Bracket.ipt** file. Use the **Find** option to locate the file if it is not displayed in the *File name* list box.

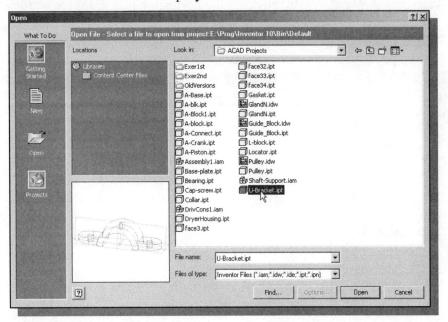

5. Click on the **Open** button in the *Startup* dialog box to accept the selected settings.

Drawing Mode – 2D Paper Space

➢ *Autodesk Inventor* allows us to generate 2D engineering drawings from solid models so that we can plot the drawings to any exact scale on paper. An engineering drawing is a tool that can be used to communicate engineering ideas/designs to manufacturing, purchasing, service, and other departments. Until now we have been working in ***model space*** to create our design in ***full size***. We can arrange our design on a two-dimensional sheet of paper so that the plotted hardcopy is exactly what we want. This two-dimensional sheet of paper is known as ***paper space*** in *AutoCAD* and *Autodesk Inventor*. We can place borders and title blocks, objects that are less critical to our design, on *paper space*. In general, each company uses a set of standards for drawing content, based on the type of product and also on established internal processes. The appearance of an engineering drawing varies depending on when, where, and for what purpose it is produced. However, the general procedure for creating an engineering drawing from a solid model is fairly well defined. In *Autodesk Inventor*, creation of 2D engineering drawings from solid models consists of four basic steps: drawing sheet formatting, creating/positioning views, annotations, and printing/plotting.

1. Click on the **drop-down arrow** next to the **New File** icon in the *Standard* toolbar area to display the available New File options.

2. Select **Drawing** from the option list.

➤ Note that a new graphics window appears on the screen. We can switch between the solid model and the drawing by clicking the corresponding graphics windows.

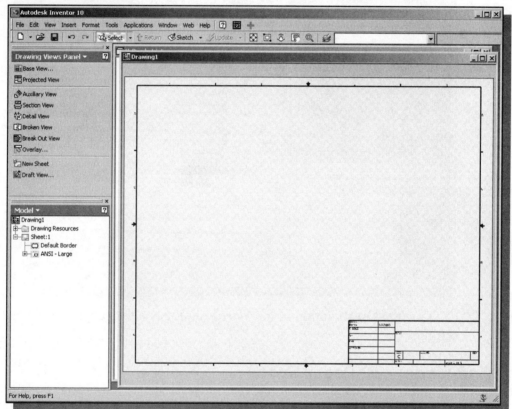

❖ In the graphics window, *Autodesk Inventor* displays a default drawing sheet that includes a title block. The drawing sheet is placed on the 2D paper space, and the title block also indicates the paper size being used.

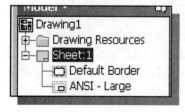

➤ In the *Browser* area, the *Drawing1* icon is displayed at the top, which indicates that we have switched to **Drawing Mode. *Sheet1*** is the current drawing sheet that is displayed in the graphics window.

Drawing Sheet Format

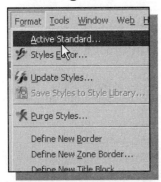

1. Choose **Format** in the pull-down menu.

2. Select **Active Standard** in the options list.

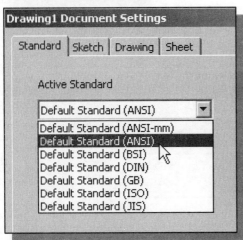

3. In the *Drawing Document Settings* dialog box, confirm that the *Current Standard* is set to **Default ANSI**.

❖ In this tutorial, we will demonstrate creating the part drawing using the default ANSI drafting standard. Note that other standards, such as ISO, GB, BSI, DIN, and JIS standrads, are also available.

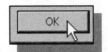

4. Click **OK** to exit the dialog box.

5. Select **Styles Editor** in the **Format** pull-down menu.

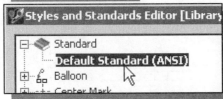

6. In the *Styles and Standards Editor* dialog box, click **Default Standard (ANSI)** to toggle the display of detailed settings for the current standard.

7. In the *General option* page, confirm that the *Projection* type is set to **Third Angle of projection**.

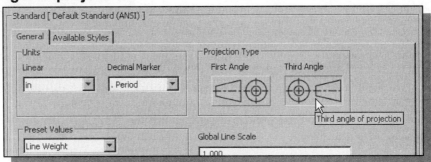

❖ Notice the different settings available in the *General option* window, such as the *Units* setting and the *Line Weight* setting.

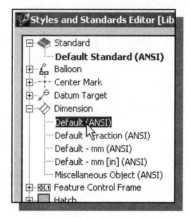

8. Choose **Default (ANSI)** in the **Dimension** list as shown.

➢ The default *Dimension Style* in Inventor is based on the ANSI Y14.5-1994 standard.

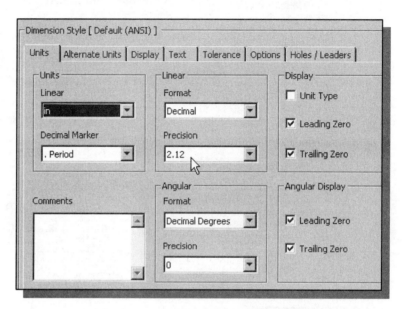

9. The **Units** tab contains the settings for linear/angular units. Note that the *Linear* units is set to **Decimal** and the precision for linear dimensions is set to two digits after the decimal point.

10. Click on the **Text** tab to display and examine the settings for dimension text. Note that the default *Dimension Style*, DEFAULT-ANSI, cannot be modified. However, new *Dimension Styles* can be created and modified.

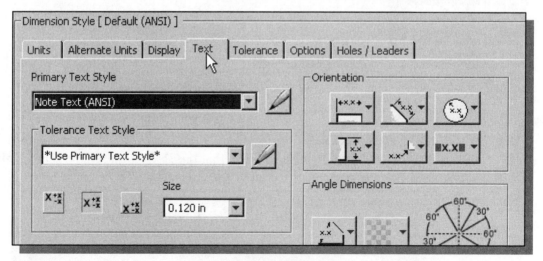

➢ On your own, display and examine the other *Settings* available.

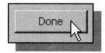

11. Click on the **Done** button to exit the *Styles and Standards Editor* dialog box.

Using the Pre-defined Drawing Sheet Formats

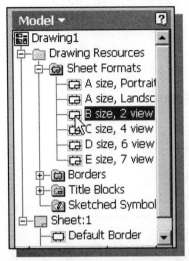

1. Inside the *Drawing Browser* window, click on the [**+**] symbol in front of **Drawing Resources** to display the available options.

2. Click on the [**+**] symbol in front of **Sheet Formats** to display the available pre-defined sheet formats.

❖ Notice several pre-defined *sheet formats*, each with a different view configuration, are available in the *Browser* window.

3. Inside the *Browser* window, double-click on the *B size, 2 view* sheet format.

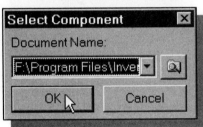

4. Click on the **OK** button to accept the default part file and generate the 2D views.

➢ The *U-Bracket* model is the only model opened. By default, all of the 2D drawings will be generated from this model file.

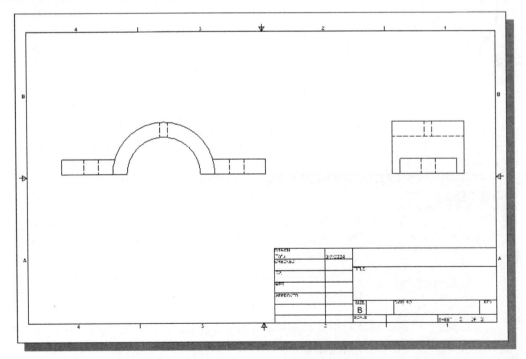

❖ We have created a B-size drawing of the *U-Bracket* model. *Autodesk Inventor* automatically generates and positions the front view and side view of the model inside the title block.

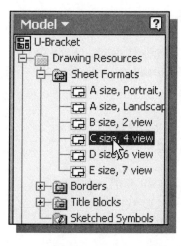

5. Inside the *Browser* window, double-click on the **C size, 4 view** sheet format.

6. Click on the **OK** button to use the default part file, *U-Bracket.ipt*, to generate the 2D views.

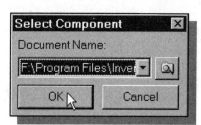

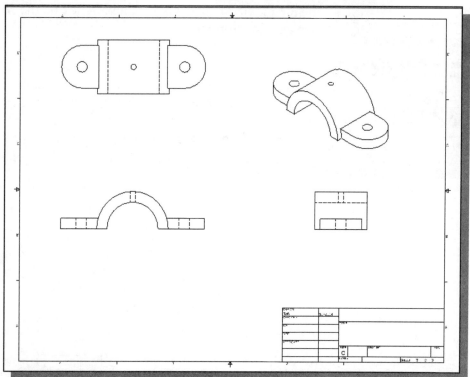

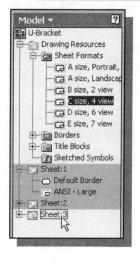

❖ Note that we have created three drawing sheets, displayed in the *Drawing Browser* window as *Sheet1*, *Sheet2*, and *Sheet3*. *Autodesk Inventor* allows us to create multiple 2D drawings from the same model file, which can be used for different purposes.

➢ In most cases, the pre-defined *sheet formats* can be used to quickly setup the views needed. However, it is also important to understand the concepts and principles involved in setting up the views. In the next sections, the procedures to set up drawing sheets and different types of views are illustrated.

Deleting, Activating, and Editing a Drawing Sheet

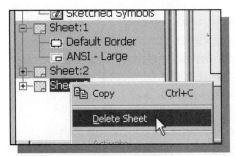

1. Inside the *Drawing Browser* window, right-mouse-click on **Sheet3** to display the option menu.

2. Select **Delete Sheet** in the option menu to remove the *Sheet3* drawing.

3. In the *warning window*, click on the **OK** button to proceed with deleting the drawing.

❖ Note that **Sheet3** is removed and **Sheet2** now becomes the active drawing sheet.

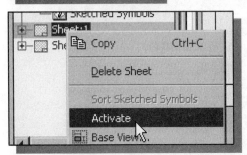

4. Inside the *Drawing Browser* window, right-mouse-click on **Sheet1** to display the option menu.

5. Select **Activate** in the option menu to set the *Sheet1* drawing as the active drawing sheet.

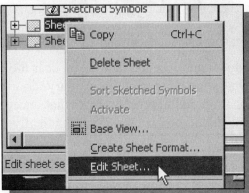

6. Inside the *Drawing Browser* window, right-mouse-click on **Sheet1** to display the option menu.

7. Select **Edit Sheet** in the option menu to display the settings for the *Sheet1* drawing.

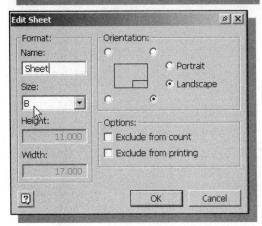

8. Set the sheet size to **B-size** and click on the **OK** button to exit the *Edit Sheet* dialog box.

Adding a Base View

❖ In *Autodesk Inventor Drawing Mode*, the first drawing view we create is called a **base view**. A *base view* is the primary view in the drawing; other views can be derived from this view. When creating a *base view*, *Autodesk Inventor* allows us to specify the view to be shown. By default, *Autodesk Inventor* will treat the *world XY plane* as the front view of the solid model. Note that there can be more than one *base view* in a drawing.

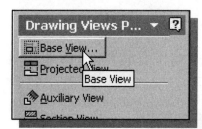

1. Click on the **Base View** in the *Drawing Views Panel* to create a base view.

2. In the *Drawing View* dialog box, confirm that the settings are set to **Front View** and **Hidden Line** as shown in the figure below. (**DO NOT** click on the **OK** button at this point.)

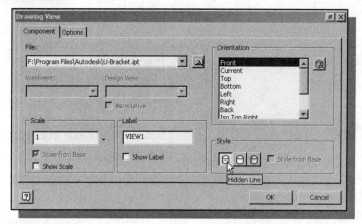

3. Move the cursor inside the graphics window and place the *base* view near the lower left corner of the graphics window as shown below. (If necessary, drag the *Create View* dialog box to another location on the screen.) Note that once the *base view* is placed, the *Drawing View* dialog box is closed automatically.

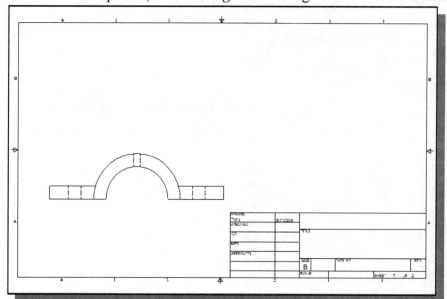

Creating Projected Views

❖ In *Autodesk Inventor Drawing Mode*, **projected views** can be created with a first-angle or third-angle projection, depending on the drafting standard used for the drawing. We must have a base view before a projected view can be created. Projected views can be orthographic projections or isometric projections. Orthographic projections are aligned to the base view and inherit the base view's scale and display settings. Isometric projections are not aligned to the base view.

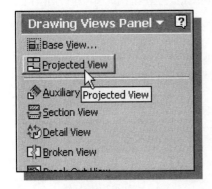

1. Click on the **Projected View** button in the *Drawing Views Panel* to create a projected view.

2. Click on the **Base View** to select the base view for the projection.

3. Move the cursor above the base view and select a location to position the projected side view of the model.

4. Move the cursor toward the upper right corner of the title block and select a location to position the isometric view of the model as shown below.

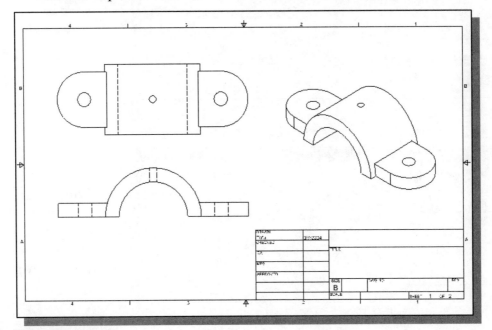

5. Inside the graphics window, right-mouse-click once to bring up the option menu.

6. Select **Create** to proceed with creating the two projected views.

Adjusting the View Scale

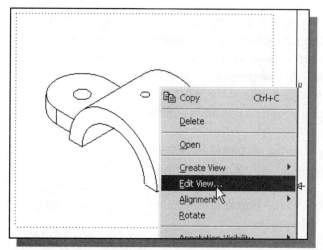

1. Move the cursor on top of the isometric view and watch for the box around the entire view indicating the view is selectable as shown in the figure. Right-mouse-click once to bring up the option menu.

2. Select **Edit View** in the option menu.

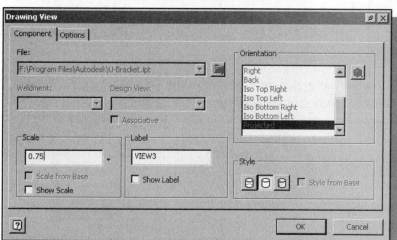

3. Inside the *Drawing View* dialog box, set the *Scale* to **0.75** as shown in the figure.

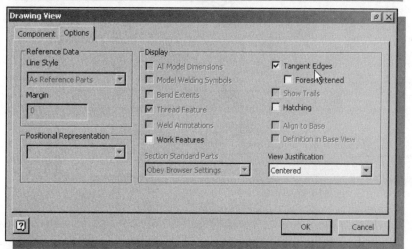

4. Click on the **Options** tab and confirm the **Tangent Edges** option is switched *ON* as shown.

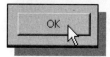

5. Click on the **OK** button to accept the settings and proceed with updating the drawing views.

Repositioning Views

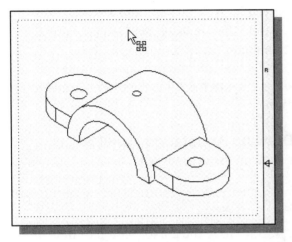

1. Move the cursor on top of the isometric view and watch for the **four-arrow Move symbol** as the cursor is near the border indicating the view can be dragged to a new location as shown in the figure.

2. Press and hold down the left-mouse-button and reposition the view to a new location.

3. On your own, reposition the views we have created so far. Note that the top view can be repositioned only in the vertical direction. The top view remains aligned to the base view, the front view.

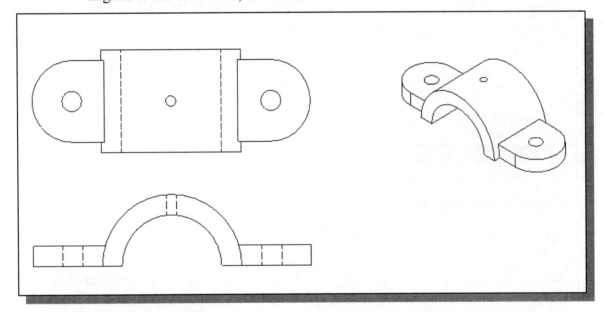

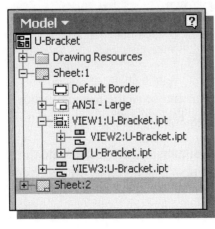

➢ Note that in the *Drawing Browser* area, a hierarchy of the created views is displayed under *Sheet1*. The base view, *View1*, is listed as the first view created, with *View2* linked to it. The top view, *View 2*, is projected from the base view, *View1*. The implied parent/child relationship is maintained by the system. Drawing views are associated with the model and the drawing sheets. As we create views from the base view, they are nested beneath the base view in the *Browser*.

Displaying Feature Dimensions

- By default, feature dimensions are not displayed in 2D views in *Autodesk Inventor*. We can change the default settings while creating the views or switch on the display of the parametric dimensions using the option menu.

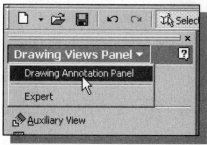

1. Left-mouse-click in the *title* area of the *Drawing Views Panel*.

2. Select **Drawing Annotation Panel** by left-clicking once in the drop-down list.

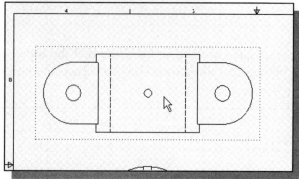

3. Move the cursor on top of the ***top*** view of the model and watch for the box around the entire view indicating the view is selectable as shown in the figure.

4. Inside the graphics window, right-mouse-click once on the top view to bring up the option menu.

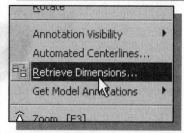

5. Select **Retrieve Dimensions** to display the parametric dimensions used to create the model.

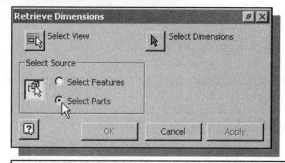

6. In the *Retrieve Dimensions* dialog box, set the *Select Source* option to **Select Parts** as shown.

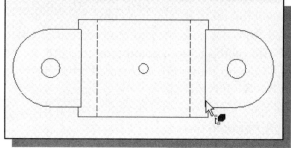

7. Move the cursor to the ***top*** view and select the *U-Bracket* part as shown.

➢ All the dimensions used to create the part are now displayed in the selected view.

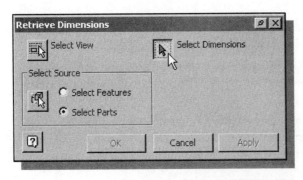

8. In the *Retrieve Dimensions* dialog box, switch on the **Select dimensions** option as shown.

➢   The system now expects us to select the dimensions to retrieve.

9. On your own, select the dimensions to retrieve by left-clicking once on the dimensions as shown. (Note that only selected dimensions are retrieved.)

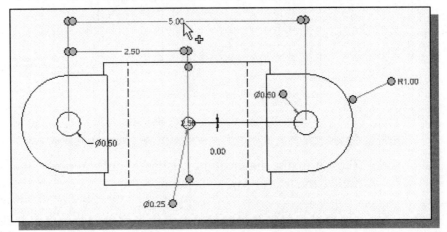

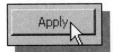

10. Click on the **Apply** button to proceed with retrieving the selected dimensions.

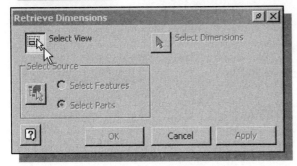

11. In the *Retrieve Dimensions* dialog box, switch on the **Select View** option as shown.

12. Select the *front* view.

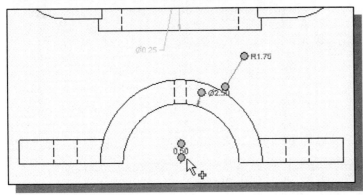

13. On your own, retrieve all three dimensions by using the **Select Dimensions** option.

14. Click on the **Cancel** button to end the Retrieve Dimensions command.

Repositioning and Hiding Feature Dimensions

1. Move the cursor on top of the width dimension text **5.00** and watch for when the dimension text becomes highlighted with the four-arrow symbol indicating the dimension is selectable.

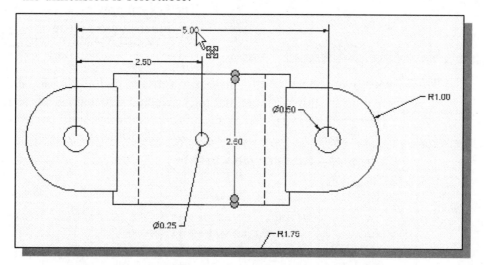

2. Reposition the dimension by using the left-mouse-button and drag the dimension text to a new location.

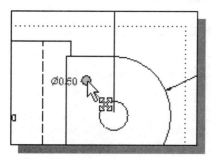

3. Move the cursor on top of the diameter dimension **0.5** and drag the grip point green dot associated with the dimension to reposition the dimension. Note that we can also drag on the dimension text, which only repositions the text.

4. On your own, reposition the dimensions displayed in the *top* view as shown in the figure below.

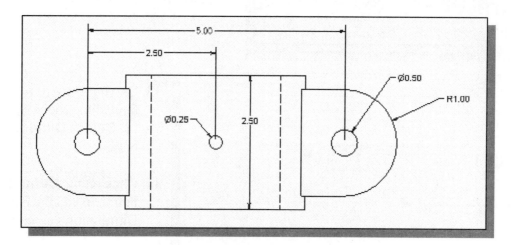

5. Move the cursor on top of the radius dimension **R 1.75** and right-mouse-click once to bring up the option menu.

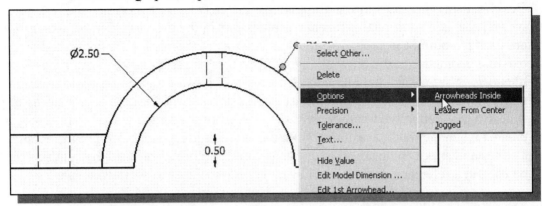

6. Select **Options → Arrowheads Inside** to display the arrowhead on the inside of the geometry.

7. On your own, reposition the dimensions as shown in the below figure.

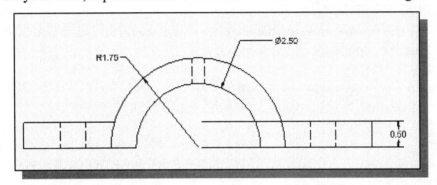

➤   Not that the displayed height dimension **0.5** is attached to the center of the design. Any feature dimensions can be removed from the display just as they are displayed.

8. Move the cursor on top of the height dimension **0.50** and right-mouse-click once to bring up the option menu.

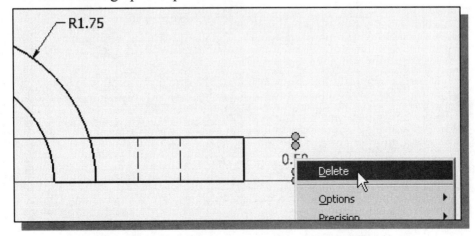

9. Select **Delete** to remove the dimension from the display.

Adding Additional Dimensions – Reference Dimensions

*   Besides displaying the **feature dimensions**, dimensions used to create the features, we can also add additional **reference dimensions** in the drawing. *Feature dimensions are used to control the geometry, whereas reference dimensions are controlled by the existing geometry. In the drawing layout, therefore, we can* ***add*** *or* ***delete*** *reference dimensions but we can only hide the feature dimensions.* One should try to use as many *feature dimensions* as possible and add *reference dimensions* only if necessary. It is also more effective to use *feature dimensions* in the drawing layout since they are created when the model was built. Note that additional *Drawing Mode* entities, such as lines and arcs, can be added to drawing views. Before *Drawing Mode* entities can be used in a reference dimension, they must be associated to a *drawing view*.

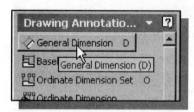

1. In the *Drawing Annotation* window, click on the **General Dimension** button.

2. In the prompt area, the message *"Select first object:"* is displayed. Select the right bottom edge of the front view.

3. In the prompt area, the message *"Select second object or place dimension:"* is displayed. Select the top horizontal edge of the front view.

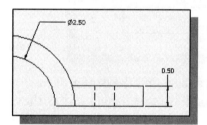

4. Pick a location that is toward the right side of the front view to place the dimension text as shown.

5. On your own, position the necessary dimensions for the design as shown in the figure below.

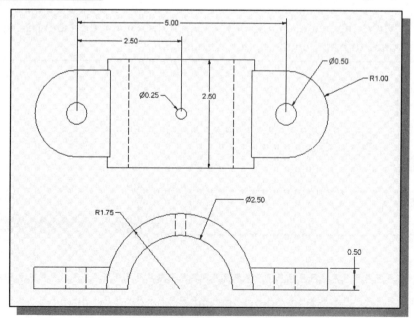

Adding Center Marks and Center Lines

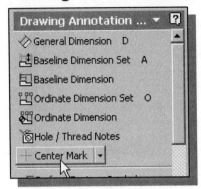

1. Click on the **Center Mark** button in the *Drawing Annotation* window.

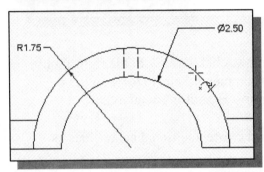

2. Click on the larger arc in the front view to add the center mark.

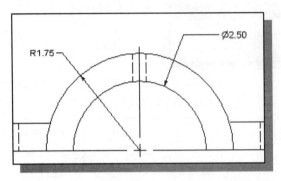

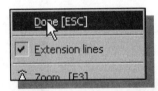

3. Inside the graphics window, click once with the right-mouse-button to display the option menu. Select **Done** in the popup menu to end the Center Mark command.

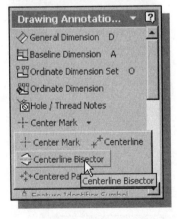

4. Click on the **drop-down arrow** next to the **Center Mark** button in the *Drawing Annotation* window to display the available options.

5. Select **Centerline Bisector** from the option list.

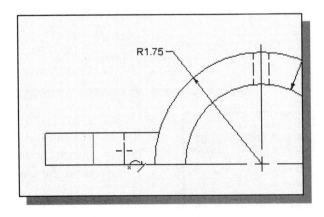

6. Inside the graphics window, click on the two hidden edges of the *drill* feature on the left as shown in the figure.

7. On your own, repeat the above step and create another centerline on the right side of the front view as shown.

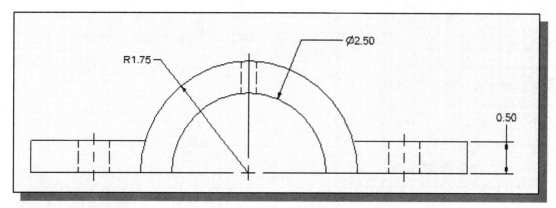

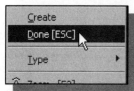

8. Inside the graphics window, click once with the right-mouse-button to display the option menu. Select **Done** in the popup menu to end the **Centerline Bisector** command.

9. On your own, repeat the above steps and create additional centerlines as shown in the figure below.

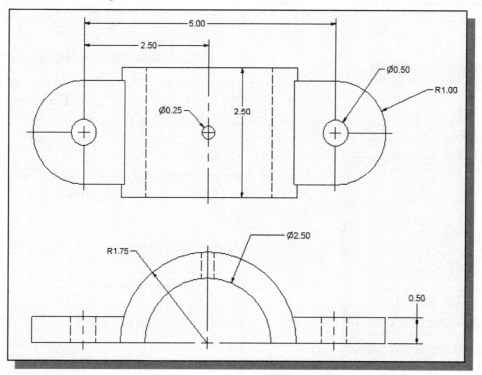

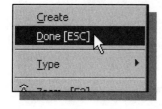

10. Inside the graphics window, click once with the right-mouse-button to display the option menu. Select **Done** in the popup menu to end the **Centerline** command.

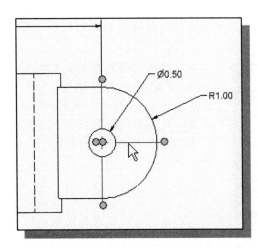

11. Click on the right centerlines in the *top* view as shown.

12. Adjust the length of the horizontal centerline by dragging on one of the grip points as shown.

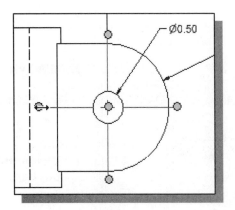

13. On your own, repeat the above steps and adjust the dimensions/centerlines as shown below.

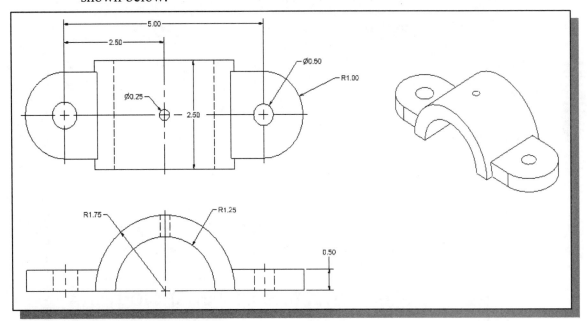

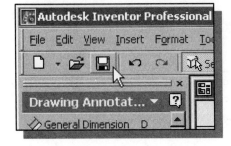

14. Click on the **Save** icon in the *Standard* toolbar as shown.

Completing the Drawing Sheet

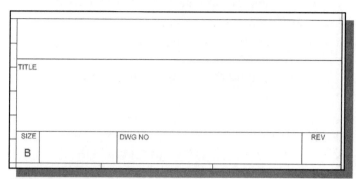

1. On your own, use the **Zoom** and **Pan** commands to adjust the display as shown, this is so that we can complete the title block.

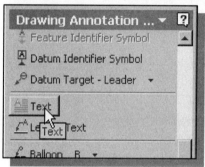

2. In the *Drawing Annotation* window, click on the **Text** button.

3. Pick a location that is inside the title block area as the location for the new text to be entered.

4. In the *Format Text* dialog box, enter the name of your organization as shown. Also note the different settings available.

5. Click **OK** to proceed.

6. On your own, repeat the above steps and complete the title block.

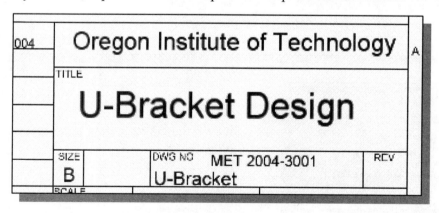

Associative Functionality – Modifying Feature Dimensions

- *Autodesk Inventor's associative functionality* allows us to change the design at any level, and the system reflects the changes at all levels automatically.

    1. Click on the **U-Bracket** part window to switch to the *Part Modeling Mode*.

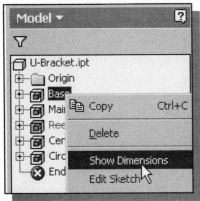

2. In the *Browser* window, right-click once on **Extrusion1** to bring up the option menu.

3. Select **Show Dimensions** in the popup option menu.

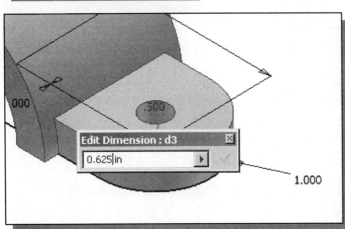

4. Double-click on one of the diameter dimensions (**0. 50**) of the drill feature on the base feature as shown in figure.

5. In the *Edit Dimension* dialog box, enter **0.625** as the new diameter dimension.

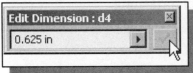

6. Click on the **check mark** button to accept the new setting.

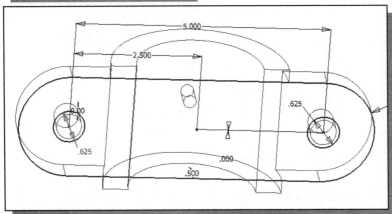

7. On your own, adjust the diameter of the other drill feature to **0.625** as shown.

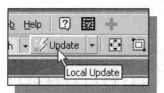

8. Click on the **Update** button in the *Standard* toolbar area to proceed with updating the solid model.

9. Click on the *U-Bracket* drawing graphics window to switch to the paper space.

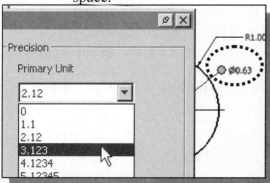

10. Inside the graphics window, double-click on the **0.63** dimension in the *top* view to bring up the *Dimension tolerance* dialog box.

11. Set the *precision* option to **3 digits after the decimal point** as shown.

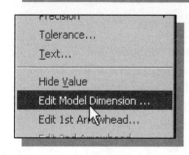

12. Inside the graphics window, right-click once on the **R 1.75** dimension in the *front* view to bring up the option menu.

13. Select **Edit Model Dimension** in the popup menu.

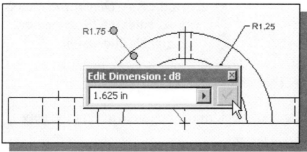

14. Change the dimension to **1.625**.

15. Click on the **check mark** button to accept the setting.

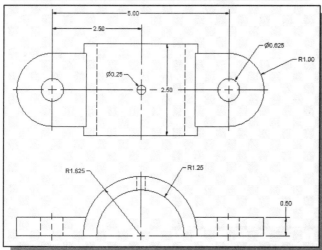

➤ Note the geometry of the cut feature is updated in all views automatically.

❖ On your own, switch to the *Part Modeling Mode* and confirm the design is updated as well.

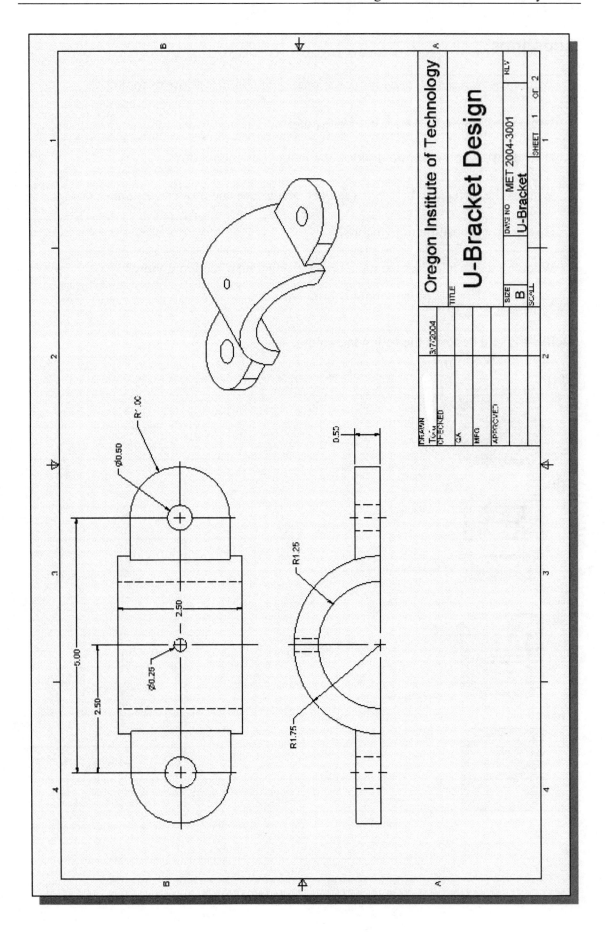

Questions:

1. What does *Autodesk Inventor*'s *associative functionality* allow us to do?

2. How do we move a view on the *Drawing Sheet*?

3. How do we display feature dimensions in the drawing mode?

4. What is the difference between a *feature dimension* and a *reference dimension*?

5. How do we reposition dimensions?

6. What are the required elements in order to generate a sectional view?

7. What is a *base view*?

8. Identify and describe the following commands:

    (a)

    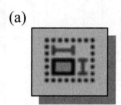

    (b)

    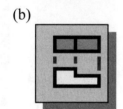

    (c)

    (d)

    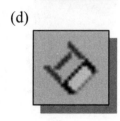

Exercises: (Create the solid models and the associated 2D drawings.)

1.   Dimensions are in inches.

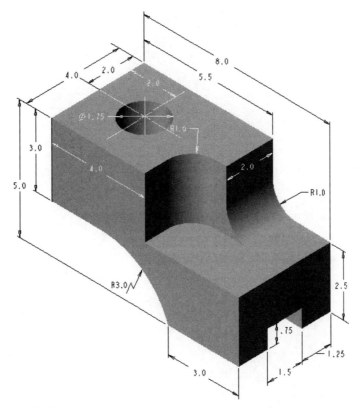

2.   Dimensions are in inches.

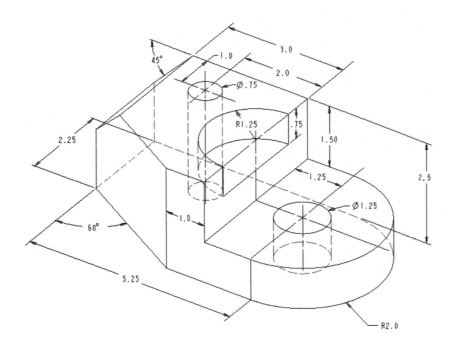

NOTES:

Chapter 9
Datum Features and Auxiliary Views

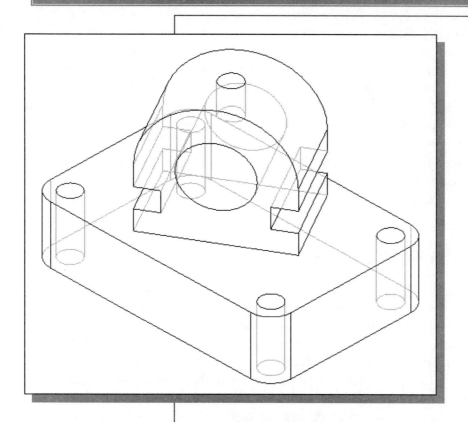

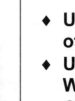

Learning Objectives

- ♦ **Understand the Concepts and the Use of Work Features**
- ♦ **Using the Different Options to Create Work Features**
- ♦ **Creating Auxiliary Views in 2D Drawing Mode**
- ♦ **Creating and Adjusting Centerlines**
- ♦ **Create Shaded Images in the 2D Drawing mode**

Work Features

Feature-based parametric modeling is a cumulative process. The relationships that we define between features determine how a feature reacts when other features are changed. Because of this interaction, certain features must, by necessity, precede others. A new feature can use previously defined features to define information such as size, shape, location and orientation. *Autodesk Inventor* provides several tools to automate this process. Work features can be thought of as user-definable datum, which are updated with the part geometry. We can create work planes, axes, or points that do not already exist. Work features can also be used to align features or to orient parts in an assembly. In this chapter, the use of the **Offset** option and the **Angled** option to create new work planes, surfaces that do not already exist, are illustrated. By creating parametric work features, the established feature interactions in the CAD database assure the capturing of the design intent. The default work features, which are aligned to the origin of the coordinate system, can be used to assist the construction of the more complex geometric features.

Auxiliary Views in 2D Drawings

An important rule concerning multiview drawings is to draw enough views to accurately describe the design. This usually requires two or three of the regular views, such as a front view, a top view and/or a side view. However, many designs have features located on inclined surfaces that are not parallel to the regular planes of projection. To truly describe the feature, the true shape of the feature must be shown using an **auxiliary view**. An *auxiliary view* has a line of sight that is perpendicular to the inclined surface, as viewed looking directly at the inclined surface. An *auxiliary view* is a supplementary view that can be constructed from any of the regular views. Using the solid model as the starting point for a design, auxiliary views can be easily created in 2D drawings. In this chapter, the general procedure of creating auxiliary views in 2D drawings from solid models is illustrated.

The *Rod-Guide* Design

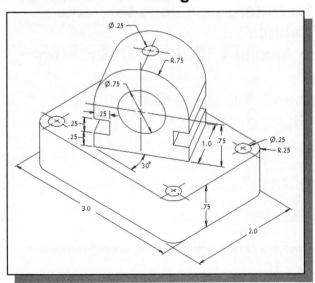

❖ Based on your knowledge of *Autodesk Inventor* so far, how would you create this design? What are the more difficult features involved in the design? Take a few minutes to consider a modeling strategy and do preliminary planning by sketching on a piece of paper. You are also encouraged to create the design on your own prior to following through the tutorial.

Modeling Strategy

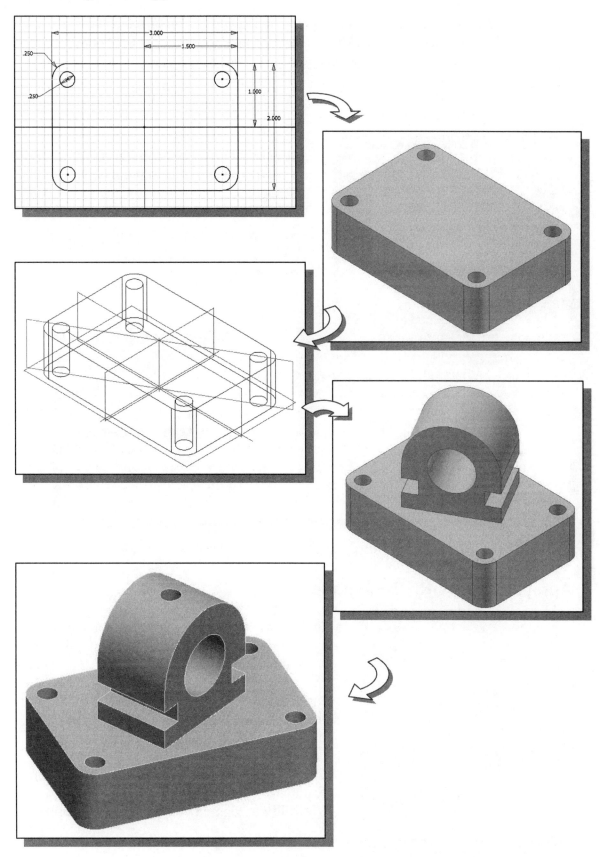

Starting Autodesk Inventor

1. Select the *Autodesk Inventor* option on the *Start* menu or select the *Autodesk Inventor* icon on the desktop to start *Autodesk Inventor*. The *Autodesk Inventor* main window will appear on the screen.

2. Once the program is loaded into the memory, the **Startup** dialog box appears at the center of the screen.

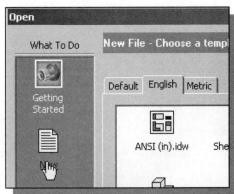

3. Select the **New** icon with a single click of the left-mouse-button in the *What to Do* dialog box.

4. Select the **English** tab and in the *New File - Choose a Template* area, select **Standard(in).ipt**.

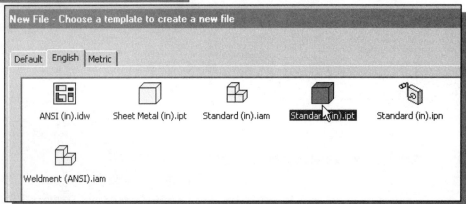

5. Pick **OK** in the *Startup* dialog box to accept the selected settings.

Applying the BORN Technique

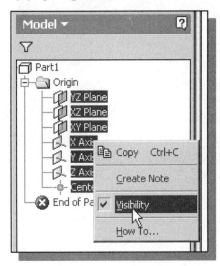

1. In the *Part Browser* window, click on the [**+**] symbol in front of the **Origin** feature to display more information on the feature.

2. Inside the *Browser* window, select all of the work features by holding down the **[Shift]** key and click with the left-mouse-button.

3. Click the right-mouse-button on any of the work features to display the option menu. Click on **Visibility** to toggle *ON* the display of the plane.

4. On your own, use the dynamic viewing options (3D Rotate, Zoom and Pan) to view the work features established.

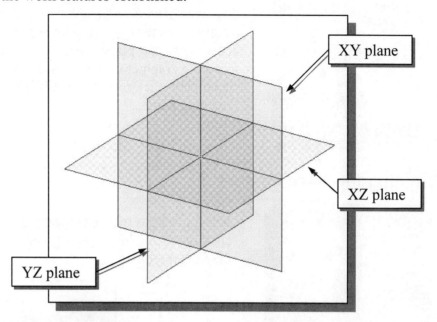

5. In the *Standard* toolbar select the **Sketch** command by left-clicking once on the icon.

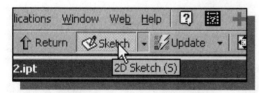

6. In the *Status Bar* area, the message: "*Select face, workplane, sketch or sketch geometry*" is displayed. *Autodesk Inventor* expects us to identify a planar surface where the 2D sketch of the next feature is to be created. Move the graphics cursor on top of *XZ Plane*, inside the *Browser* window as shown, and notice that *Autodesk Inventor* will automatically highlight the corresponding plane in the graphics window. Left-click once to select the *XZ Plane* as the sketching plane.

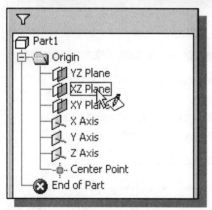

7. Select the **Project Geometry** command in the *2D Sketch Panel*. The Project Geometry command allows us to project existing features to the active sketching plane. Left-click once on the icon to activate the Project Geometry command.

8. In the *Status Bar* area, the message: "*Select edge, vertex, work geometry or sketch geometry to project*" is displayed. *Autodesk Inventor* expects us to select any existing geometry which will be projected onto the active sketching plane. Inside the *Browser* window, move the graphics cursor on top of **Center Point** as shown, left-click once to project the center point onto the sketching plane.

Creating the Base Feature

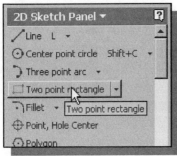
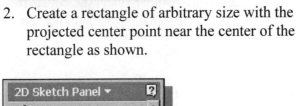

1. Select the **Two point rectangle** command by clicking once with the left-mouse-button on the icon in the *Sketch* toolbar.

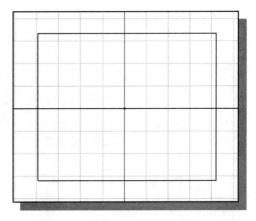

2. Create a rectangle of arbitrary size with the projected center point near the center of the rectangle as shown.

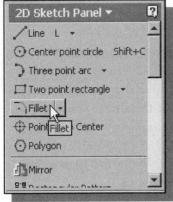

3. Click on the **Fillet** icon in the *2D Sketch Panel*.

4. The *2D Fillet* radius dialog box appears on the screen. Use the default radius value and create four rounded corners as shown.

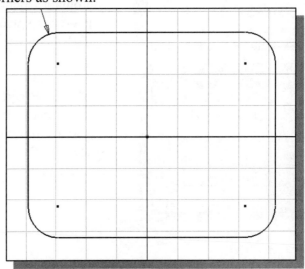

5. On your own, create four circles of the same diameter and with the centers aligned to the centers of arcs. Also create and modify the six dimensions as shown in the below figure.

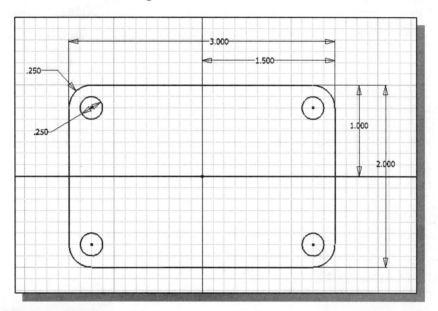

6. Inside the graphics window, click once with the **right-mouse-button** and select **Finish Sketch** in the popup menu to end the Sketch option.

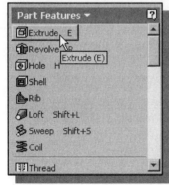

7. In the *Part Features* toolbar, select the **Extrude** command by releasing the left-mouse-button on the icon.

8. Select the inside region of the 2D sketch to create a profile as shown.

9. In the *Extrude* popup window, enter **0.75** as the extrusion distance.

10. Click on the **OK** button to proceed with creating the feature.

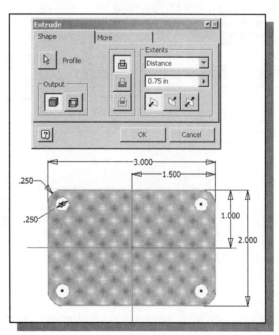

Creating an Angled Work Plane

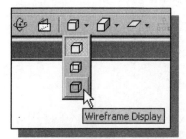

1. In the *Standard* toolbar, select the **Wireframe Display** command to set the display mode to wireframe.

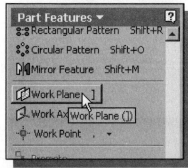

2. In the *Part Features* toolbar, select the **Work Plane** command by left-clicking the icon.

3. In the *Status Bar* area, the message: "*Define work plane by highlighting and selecting geometry*" is displayed. *Autodesk Inventor* expects us to select any existing geometry, which will be used as a reference to create the new work plane.

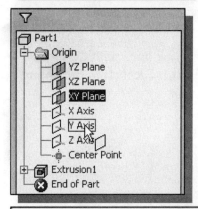

4. Inside the *Browser* window, left-click once to select the *XY Plane* as the first reference of the new work plane.

5. Inside the *Browser* window, left-click once to select the *Y Axis* as the second reference of the new work plane.

6. In the *Angle* popup window, enter **30** as the rotation angle for the new work plane.

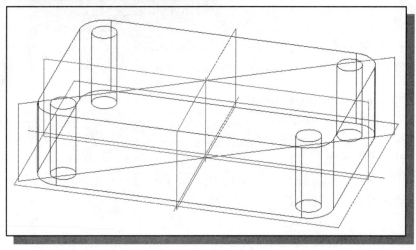

❖ Note that the *angle* is measured relative to the selected reference plane, *XY Plane*.

7. Click on the **check mark** button to accept the setting.

Creating a 2D Sketch on the Work Plane

1. In the *Standard* toolbar select the **Sketch** command by left-clicking once on the icon.

2. In the *Status Bar* area, the message: "*Select face, work plane, sketch or sketch geometry*" is displayed. Pick the work plane by clicking the work plane name inside the *Browser* as shown below.

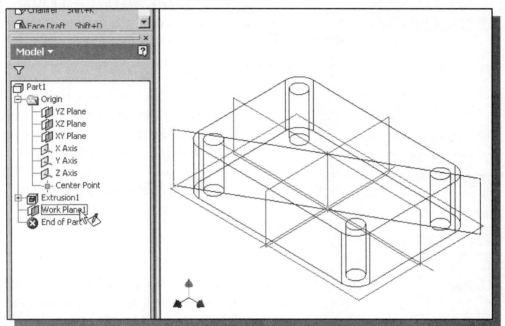

Using the Projected Geometry Option

❖ **Projected geometry** is another type of *reference geometry*. The Project Geometry tool can be used to project geometry from previously defined sketches or features onto the sketch plane. The position of the projected geometry is fixed to the feature from which it was projected. We can use the Project Geometry tool to project geometry from a sketch or feature onto the active sketch plane.

Typical uses of projected geometry include:
- Project a silhouette of a 3D feature onto the sketch plane for use in a 2D profile.
- Project the default center point onto the sketch plane to constrain a sketch to the origin of the coordinate system.
- Project a sketch from a feature onto the sketch plane so that the projected sketch can be used to constrain a new sketch.

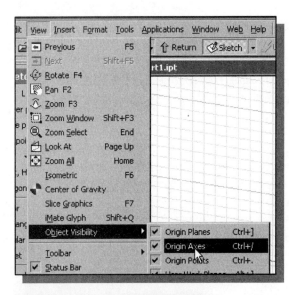

3. Choose **View** in the pull-down menu.

4. Select **Object Visibility → Origin Axes** in the options list.

➤ Note the quick-key options available to toggle *ON/OFF* the different reference geometry.

5. Use the **Ctrl+]** key combination to toggle *OFF* the display of the origin planes.

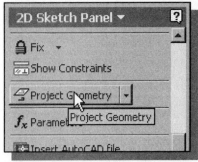

6. Select the **Project Geometry** command in the *2D Sketch Panel*. The Project Geometry command allows us to project existing features to the active sketching plane. Left-click once on the icon to activate the Project Geometry command.

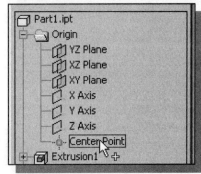

7. In the *Status Bar* area, the message: "*Select edge, vertex, work geometry or sketch geometry to project*" is displayed. Inside the *Browser* window, left-click once on the *Center Point* to project the center point onto the sketching plane.

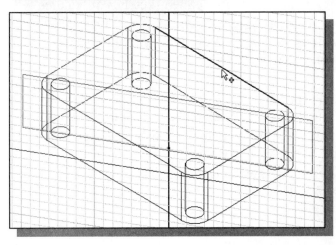

8. Select the top back edge of the base feature to create a projected line on the sketching plane.

➤ The projected line will be used in the 2D profile of the next solid feature.

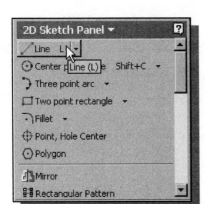

9. In the *Sketch* toolbar, click on the **Line** icon with the left-mouse-button to activate the Line command.

10. Create a rough sketch using the projected edge as the bottom line as shown in the figure. (Note that all edges are either horizontal or vertical.)

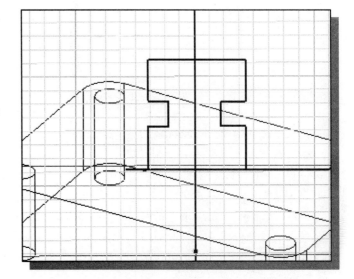

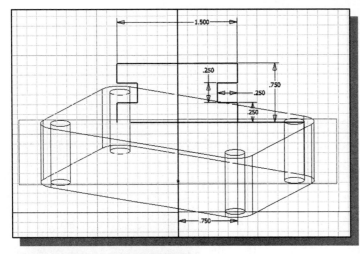

11. On your own, create and modify the dimensions as shown; note that the sketch is symmetrical vertically.

➢ Note that the projected line is used only as a reference; the gap at the bottom of the 2D sketch indicates the 2D sketch cannot be used as a profile yet.

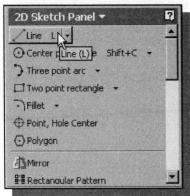

12. Select the **Line** command in the *2D Sketch Panel*.

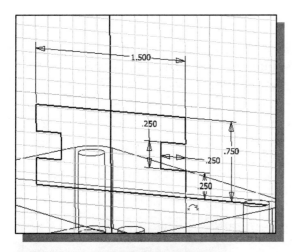

13. Create a line connecting the bottom two corners of the 2D sketch as shown in the figure.

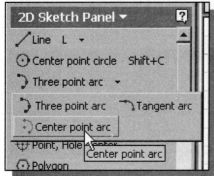

14. Select the **Center point arc** option in the *2D Sketch Panel* as shown.

15. On your own, create the arc aligned to the mid-point of the top edge as shown.

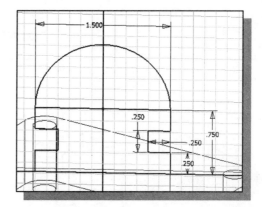

16. On your own, add a 0.75 circle and complete the 2D sketch as shown in the figure.

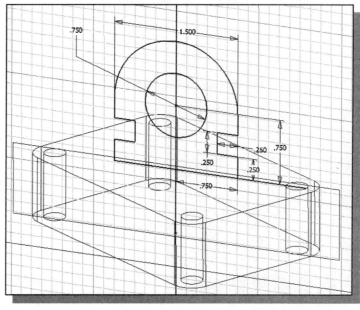

Completing the Solid Feature

1. Inside the graphics window, click once with the right-mouse-button and select **Finish Sketch** in the popup menu to end the Sketch option.

2. In the *Part Features* toolbar, select the **Extrude** command by releasing the left-mouse-button on the icon.

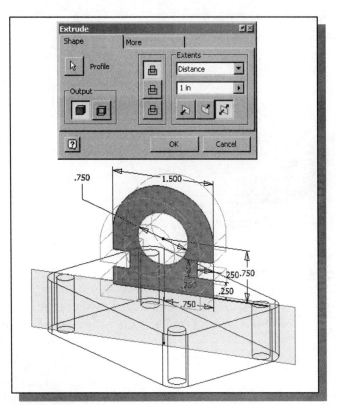

3. Select the inside region of the 2D sketch to create a profile as shown.

4. In the *Extrude* popup window, enter **1.0** as the extrusion distance.

5. Set the extrusion direction to **Both Sides** as shown.

6. Click on the **OK** button to proceed with creating the feature.

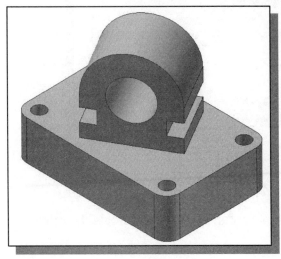

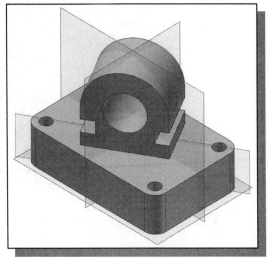

Creating an Offset Work Plane

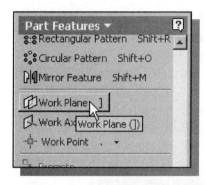

1. In the *Part Features* toolbar, select the **Work Plane** command by left-clicking the icon.

❖ In the *Status Bar* area, the message: "*Define work plane by highlighting and selecting geometry*" is displayed. *Autodesk Inventor* expects us to select any existing geometry, which will be used as a reference to create the new work plane.

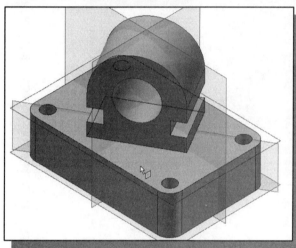

2. Inside the *graphics* window, select the *top plane* of the base feature as the reference of the new work plane.

3. Click and drag the reference plan upward until the value in the *Offset* popup window reads **0.75** as the offset distance for the new work plane.

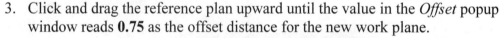

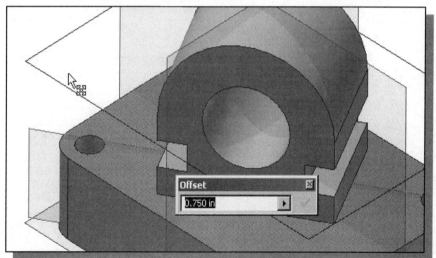

4. Click on the **check mark** button to accept the setting and create the reference plane.

Creating another cut feature using the work plane

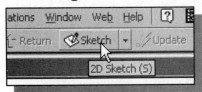

1. In the *Standard* toolbar select the **Sketch** command by left-clicking once on the icon.

2. In the *Status Bar* area, the message: "*Select face, work plane, sketch or sketch geometry*" is displayed. Pick the work plane we just created by clicking on one of the edges of the work plane as shown below.

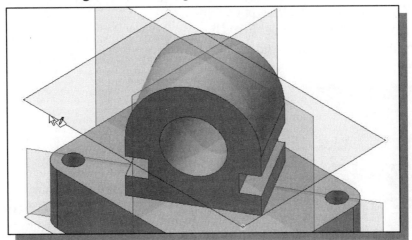

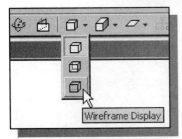

3. In the *Standard* toolbar, select the **Wireframe Display** command to set the display mode to wireframe.

4. Use the **Ctrl+]** quick-key combination to toggle off the display of the *Origin Planes*.

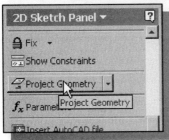

5. Select the **Project Geometry** command in the *2D Sketch Panel*. The Project Geometry command allows us to project existing features to the active sketching plane. Left-click once on the icon to activate the Project Geometry command.

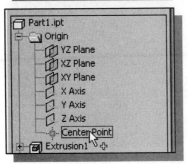

6. In the *Status Bar* area, the message: "*Select edge, vertex, work geometry or sketch geometry to project*" is displayed. Inside the *Browser* window, left-click once on the **Center Point** to project the center point onto the sketching plane.

7. Select the **Center point circle** command by clicking once with the left-mouse-button on the icon in the *Sketch* toolbar.

8. On your own, create a **Ø0.25 circle** aligned to the projected *center point* as shown in the figure below.

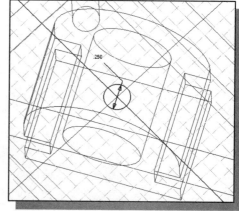

9. Inside the graphics window, click once with the right-mouse-button and select **Finish Sketch** in the popup menu to end the Sketch option.

10. In the *Part Features* toolbar, select the **Extrude** command by releasing the left-mouse-button on the icon.

11. On your own, complete the cut feature as shown.

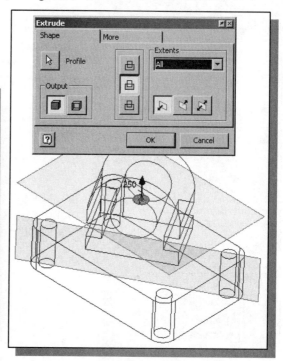

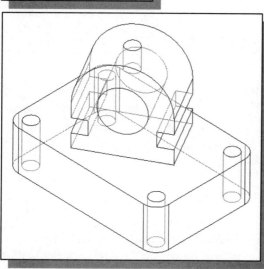

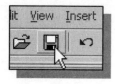

12. On your own, save the design as **Rod-Guide.ipt**.

Starting a New 2D Drawing

1. Select **New File** → **New Drawing** in the *Standard* toolbar as shown.

➢ Note that a new graphics window appears on the screen. We can switch between the solid model and the drawing by clicking the corresponding graphics windows.

❖ In the graphics window, *Autodesk Inventor* displays a default drawing sheet that includes a title block. The drawing sheet is placed on the 2D paper space, and the title block also indicates the paper size being used.

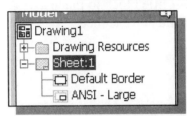

➢ In the *Browser* area, the *Drawing1* icon is displayed at the top, which indicates that we have switched to **Drawing Mode**. *Sheet:1* is the current drawing sheet that is displayed in the graphics window.

❖ Different types of pre-defined borders and title blocks are available in *Autodesk Inventor*.

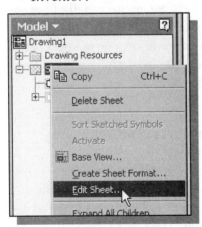

2. In the *Browser* area, click once with the right-mouse-button on *Sheet:1* and select **Edit Sheet** in the popup menu.

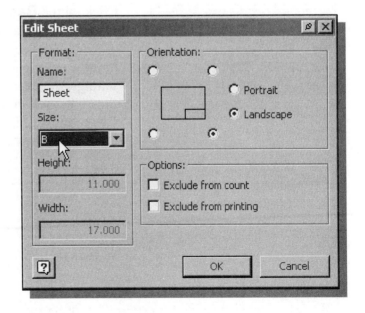

3. In the *Edit Sheet* window, set the size option to **B** size as shown.

4. Click on the **OK** button to accept the settings.

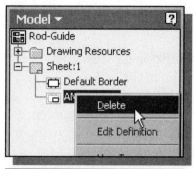

5. In the *Browser* area, click once with the right-mouse-button on ***ANSI-Large*** and select **Delete** in the popup menu.

❖ Before applying a different title block, the existing title block must be removed.

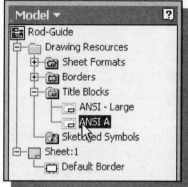

6. In the *Browser* area, click on the [+] symbol in front of the ***Drawing Resources*** and ***Title Blocks*** options to expand the lists.

7. Double-click on the ***ANSI-A*** title block to place a copy of this title block into the current sheet.

Adding a Base View

❖ In *Autodesk Inventor Drawing Mode*, the first drawing view we create is called a **base view**. A *base view* is the primary view in the drawing; other views can be derived from this view. When creating a *base view*, *Autodesk Inventor* allows us to specify the view to be shown. By default, *Autodesk Inventor* will treat the *world XY plane* as the front view of the solid model. Note that there can be more than one *base view* in a drawing.

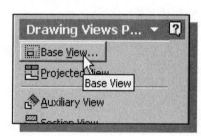

1. Click on the **Base View** in the *Drawing Views Panel* to create a base view.

2. In the *Drawing View* dialog box, confirm that the settings are set to **Top** view and **Hidden Line** as shown in the figure below. (**DO NOT** click on the **OK** button at this point.)

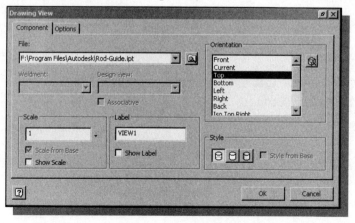

3. Move the cursor inside the graphics window and place the **base view** near the upper left corner of the graphics window as shown below. (If necessary, drag the *Create View* dialog box to another location on the screen.) Note that once the *base view* is placed, the *Drawing View* dialog box is closed automatically.

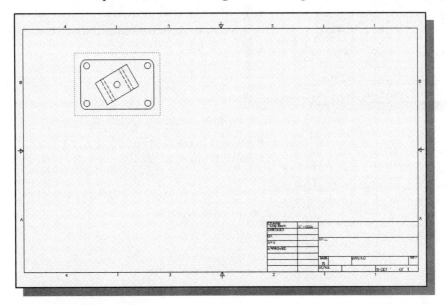

Creating an Auxiliary View

❖ In *Autodesk Inventor Drawing Mode*, the **Projected View** command is used to create standard views such as the *top* view, *front* view or *isometric* view. For non-standard views, the **Auxiliary View** command is used. *Auxiliary* views are created using orthographic projections. Orthographic projections are aligned to the base view and inherit the base view's scale and display settings.

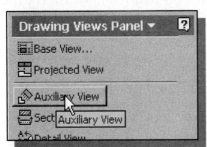

1. Click on the **Auxiliary View** button in the *Drawing Views Panel*.

2. Click on the **base view** to select the base view for the projection.

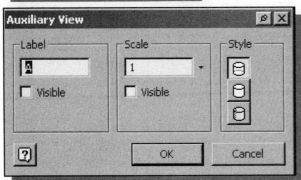

3. Confirm the settings in the *Auxiliary View* window are set as shown. (**DO NOT** click on the **OK** button.)

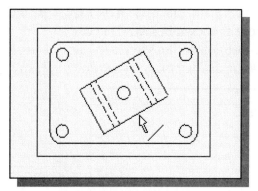

4. Pick the front edge of the upper section of the model as shown.

❖ The orthographic projection direction will be perpendicular to the selected edge.

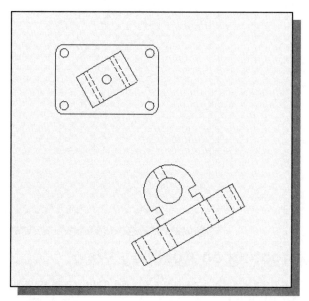

5. Move the cursor below the base view and select a location to position the auxiliary view of the model as shown.

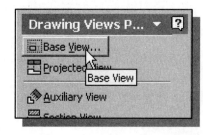

6. Click on **Base View** in the *Drawing Views Panel* to create a base view.

7. In the *Drawung View* dialog box, confirm that the settings are set to **Iso Top Right** and **Hidden Line Removed** as shown in the figure below. (**DO NOT** click on the **OK** button at this point.)

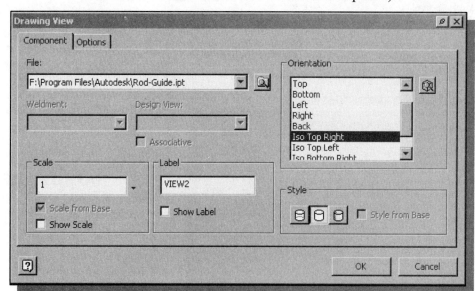

8. Move the cursor toward the upper right corner of the title block and select a location to position the *isometric* view of the model as shown below.

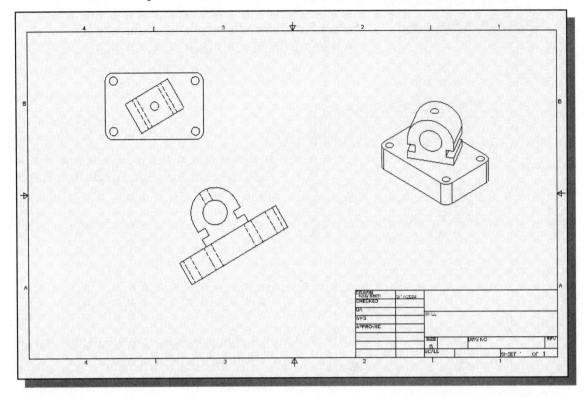

Displaying Feature Dimensions

- By default, feature dimensions are not displayed in 2D views in *Autodesk Inventor*. We can change the default settings while creating the views or switch on the display of the parametric dimensions using the option menu.

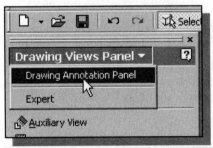

1. Left-mouse-click in the *title* area of the *Drawing Views Panel*.

2. Select **Drawing Annotation Panel** by left-clicking once in the drop-down list.

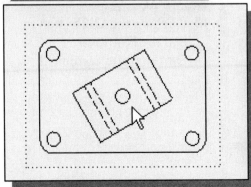

3. Move the cursor on top of the *top* view of the model and watch for the box around the entire view indicating the view is selectable as shown in the figure.

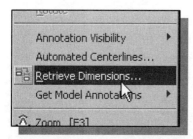

4. Inside the graphics window, right-mouse-click once to bring up the option menu.

5. Select **Retrieve Dimensions** to display the parametric dimensions used to create the model.

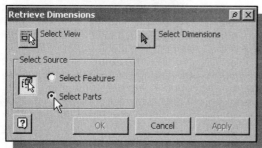

6. In the *Retrieve Dimensions* dialog box, set the *Select Source* option to **Select Parts** as shown.

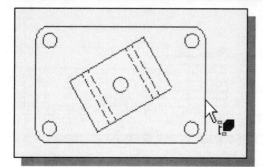

7. Move the cursor to the *top* view and select any edge of the *Rod-Guide* part as shown.

➢ All the dimensions used to create the part are now displayed in the selected view.

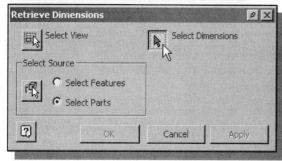

8. In the *Retrieve Dimensions* dialog box, switch on the **Select dimensions** option as shown.

➢ The system now expects us to select the dimensions to be retrieved.

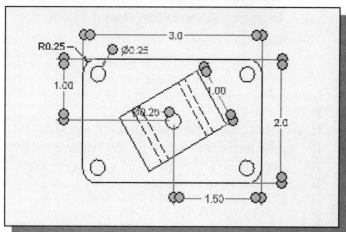

9. On your own, select all the dimensions except the radius of the round as shown.

10. Click on the **OK** button to end the Retrieve Dimensions command.

Adjusting the View Scale

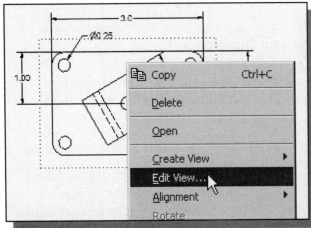

1. Move the cursor on top of the isometric view and watch for the box around the entire view indicating the view is selectable as shown in the figure. Right-mouse-click once to bring up the option menu.

2. Select **Edit View** in the option menu.

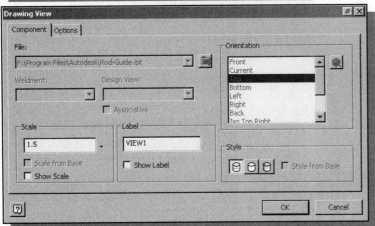

3. Inside the *Drawing View* dialog box, set the *Scale* to **1.5** as shown in the figure.

4. Click on the **OK** button to accept the settings.

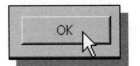

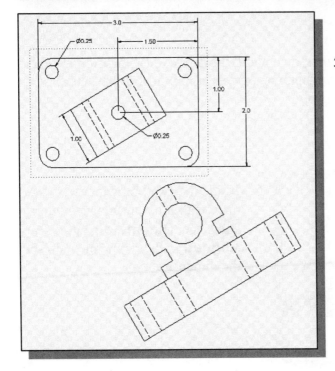

5. On your own, reposition the views and dimensions clicking and dragging the individual entities.

❖ Note that in parametric modeling software, dimensions are always associated with the geometry, even in the 2D drawing mode.

Adding Additional Dimensions

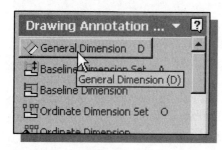

1. In the *Drawing Annotation* window, click on the **General Dimension** button.

2. In the prompt area, the message "*Select first object:*" is displayed. Select the right bottom edge of the base feature in the auxiliary view.

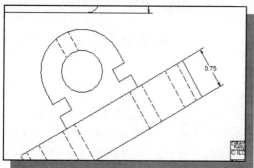

3. In the prompt area, the message "*Select second object or place dimension:*" is displayed. Select the right vertical edge of the base feature in the auxiliary view.

4. Pick a location that is toward the right side of the auxiliary view to place the dimension text as shown.

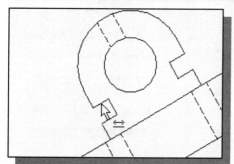

5. Select the top edge of the notch in the auxiliary view as shown.

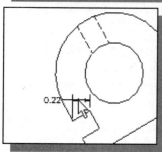

❖ Note that, by default, the displayed dimension is either the horizontal or vertical component of the actual dimension.

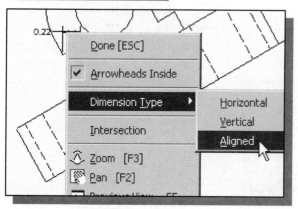

6. Inside the graphics window, click once with the right-mouse-button and select **Dimension Type → Aligned** in the popup menu to set the dimension type.

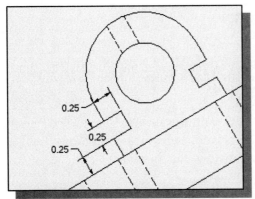

7. Create the three dimensions as shown.

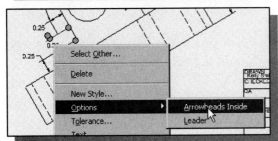

❖ Use the **Options** menu to set the arrowheads to the inside.

➢ Note that the different grip points can also be used to adjust the dimensions. You are encouraged to experiment with dragging the different parts of the dimensions and understand how to control the displayed dimensions.

8. On your own, create and position the necessary dimensions for the design as shown in the figure below.

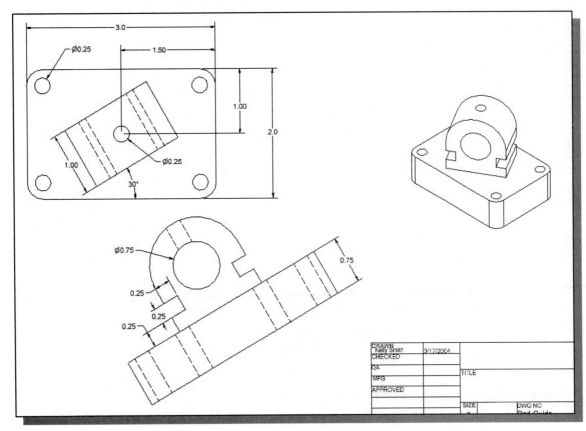

Adding Center Marks and Center Lines

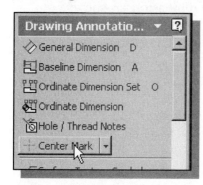

1. Click on the **Center Mark** button in the *Drawing Annotation* window.

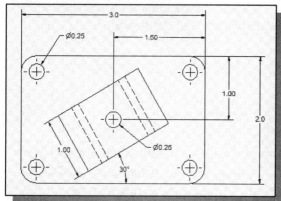

2. Click on the five circles in the top view to add the center marks as shown.

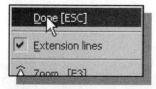

3. Inside the graphics window, click once with the right-mouse-button to display the option menu. Select **Done** in the popup menu to end the Center Mark command.

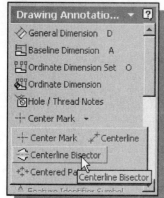

4. Click on the **drop-down arrow** next to the **Center Mark** button in the *Drawing Annotation* window to display the available options.

5. Select **Centerline Bisector** from the option list.

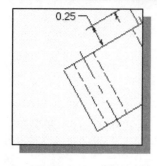

6. Inside the graphics window, click on the two hidden edges of one of the *drill* features and create a center line in the auxiliary view as shown in the figure.

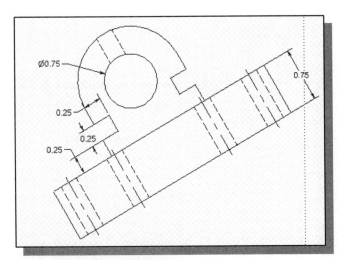

7. On your own, repeat the above step and create additional centerlines as shown.

8. Inside the graphics window, click once with the right-mouse-button to display the option menu. Select **Done** in the popup menu to end the Centerline Bisector command.

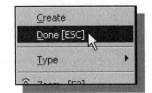

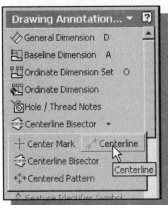

9. Click on the **drop-down arrow** next to the **Center Mark** button in the *Drawing Annotation* window to display the available options.

10. Select **Centerline** from the option list.

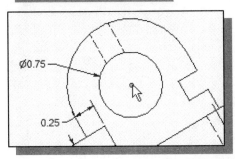

11. Click on the center point of the circle in the auxiliary view as show.

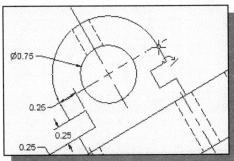

12. Click on one of the arc endpoints when the green dot appears as show.

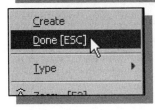

13. Inside the graphics window, click once with the right-mouse-button to display the option menu. Select **Done** in the popup menu to end the Centerline command.

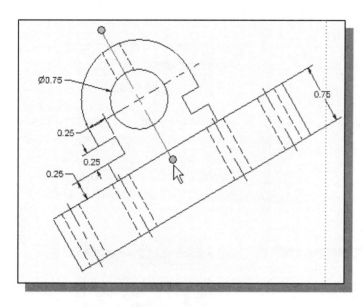

14. Click on the **centerlines** in the auxiliary view as shown.

15. Adjust the length of the horizontal centerline by dragging on one of the grip points as shown.

16. On your own, repeat the above steps and adjust the dimensions/centerlines as shown below.

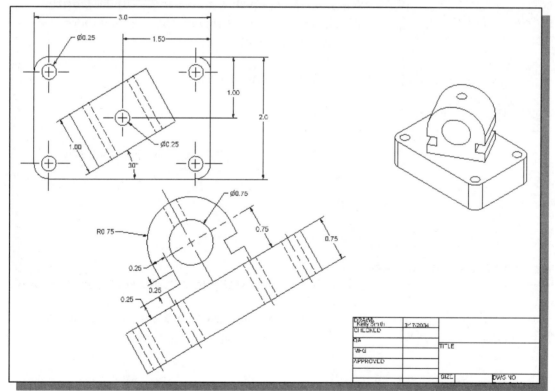

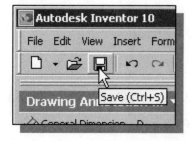

17. Click on the **Save** icon in the *Standard* toolbar and save the drawing as *Rod-Guide.idw*.

Completing the Drawing Sheet

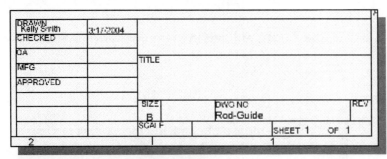

1. On your own, use the **Zoom** and **Pan** commands to adjust the display to work on the title block area.

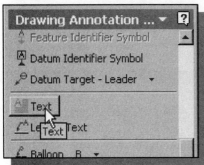

2. In the *Drawing Annotation* window, click on the **Text** button.

3. Pick a location that is inside the title block area as the location for the new text to be entered.

4. In the *Format Text* dialog box, enter the name of your organization as shown. Also note the different settings available.

5. Click **OK** to proceed.

6. On your own, repeat the above steps and complete the title block.

7. On your own, also create a general note at the lower left corner of the border as shown.

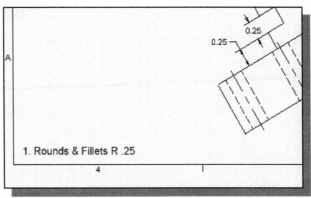

Editing the Isometric view

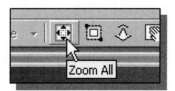

1. Click on the **Zoom All** button in the *Standard* toolbar.

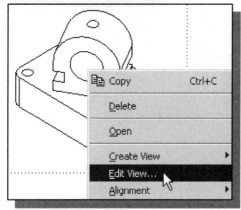

2. Right-mouse-click inside the *isometric* view to bring up the option menu as shown.

3. Select **Edit View** in the option menu.

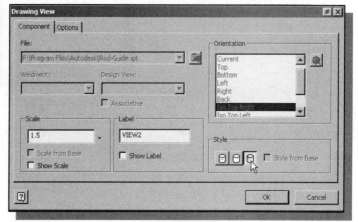

4. In the *Drawing View* dialog box, set the scale to **1.5** and the display *Style* to **Shaded**.

5. Click on the **OK** button to accept the settings.

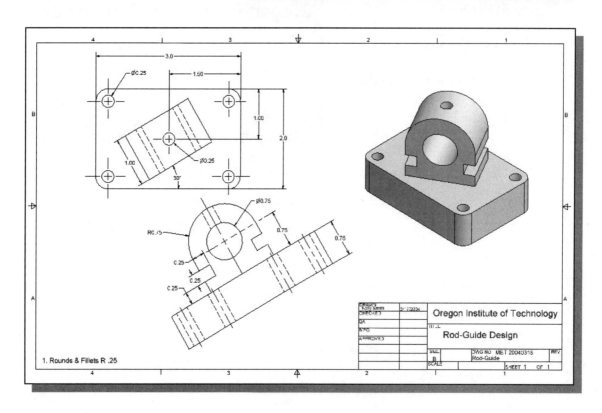

Questions:

1. What are the different types of work features available in *Autodesk Inventor*?

2. Why work features are important in parametric modeling?

3. Describe the purpose of auxiliary views in 2D drawings?

4. What are the required elements in order to generate an auxiliary view?

5. Can we use a different Title Block in the Drawing Mode? How?

6. Describe the different methods used to create centerlines in the chapter.

7. Can we change the View Scale of existing views? How?

8. Identify and describe the following commands:

(a)

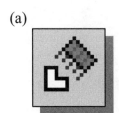

(b)

(c)

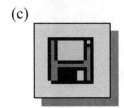

(d)

Exercises: (Create the solid models and the associated 2D drawings.)

1.   Dimensions are in inches.

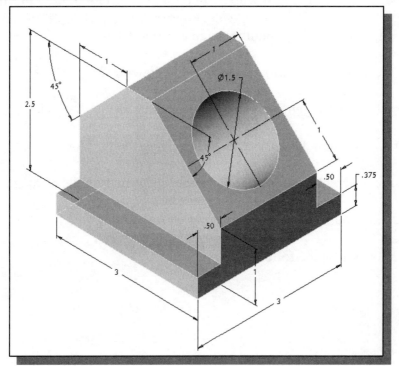

2.   Dimensions are in millimeters.

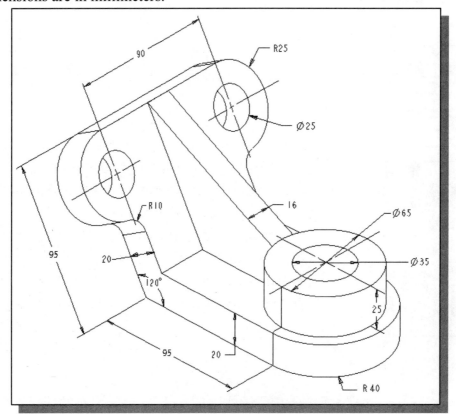

Chapter 10
Symmetrical Features in Designs

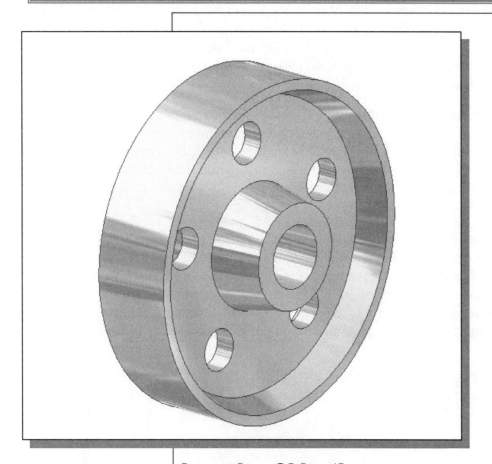

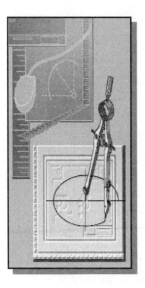

Learning Objectives

- ◆ **Create Revolved Features**
- ◆ **Use the Mirror Feature Command**
- ◆ **Create New Borders and Title Blocks**
- ◆ **Create Circular Patterns**
- ◆ **Create and Modify Linear Dimensions**
- ◆ **Use Autodesk Inventor's Associative Functionality**
- ◆ **Identify Symmetrical Features in Designs**

Introduction

In parametric modeling, it is important to identify and determine the features that exist in the design. *Feature-based parametric modeling* enables us to build complex designs by working on smaller and simpler units. This approach simplifies the modeling process and allows us to concentrate on the characteristics of the design. Symmetry is an important characteristic that is often seen in designs. Symmetrical features can be easily accomplished by the assortments of tools that are available in feature-based modeling systems, such as *Autodesk Inventor*.

The modeling technique of extruding two-dimensional sketches along a straight line to form three-dimensional features, as illustrated in the previous chapters, is an effective way to construct solid models. For designs that involve cylindrical shapes, shapes that are symmetrical about an axis, revolving two-dimensional sketches about an axis can form the needed three-dimensional features. In solid modeling, this type of feature is called a *revolved feature*.

In *Autodesk Inventor*, besides using the **Revolve** command to create revolved features, several options are also available to handle symmetrical features. For example, we can create multiple identical copies of symmetrical features with the **Feature Pattern** command, or create mirror images of models using the **Mirror Feature** command. We can also use *construction geometry* to assist the construction of more complex features. In this lesson, the construction and modeling techniques of these more advanced options are illustrated.

A Revolved Design: *PULLEY*

❖ Based on your knowledge of *Autodesk Inventor*, how many features would you use to create the design? Which feature would you choose as the **base feature** of the model? Identify the symmetrical features in the design and consider other possibilities in creating the design. You are encouraged to create the model on your own prior to following through the tutorial.

Modeling Strategy – A Revolved Design

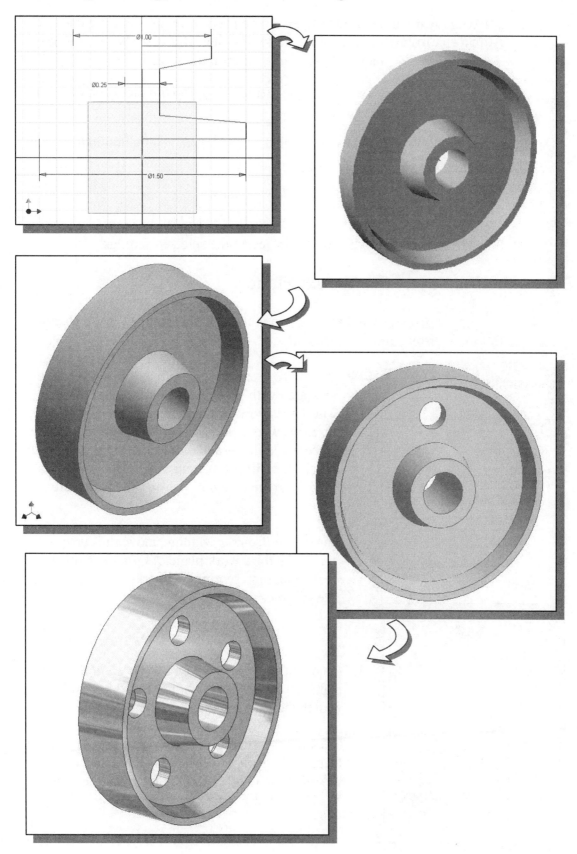

Starting Autodesk Inventor

1. Select the **Autodesk Inventor** option on the *Start* menu or select the **Autodesk Inventor** icon on the desktop to start *Autodesk Inventor*. The *Autodesk Inventor* main window will appear on the screen.

2. Once the program is loaded into the memory, the *Startup* dialog box appears at the center of the screen.

3. Select the **New** icon with a single click of the left-mouse-button in the *What to Do* dialog box.

4. Select the **English** tab and in the *New - Choose Template* area, select **Standard(in).ipt**.

5. Pick **OK** in the *Startup* dialog box to accept the selected settings.

Set up the Display of the Sketch Plane

1. In the *Part Browser* window, click on the [**+**] symbol in front of the ***Origin*** feature to display more information on the feature.

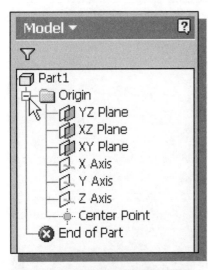

❖ In the *Browser* window, notice a new part name appeared with seven work features established. The seven work features include three *workplanes*, three *work axes* and a *work point*. By default, the three work planes and work axes are aligned to the **world coordinate system** and the work point is aligned to the *origin* of the **world coordinate system**.

2. Inside the *Browser* window, move the cursor on top of the third work plane, ***XY Plane***. Notice a rectangle, representing the work plane, appears in the graphics window.

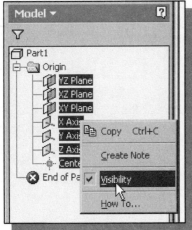

3. Inside the *Browser* window, select all of the work features by holding down the **[Shift]** key and click with the left-mouse-button.

4. Click the right-mouse-button on any of the work features to display the option menu. Click on **Visibility** to toggle *ON* the display of the plane.

Creating the 2D Sketch for the Base feature

1. In the *Standard* toolbar select the **Sketch** command by left-clicking once on the icon.

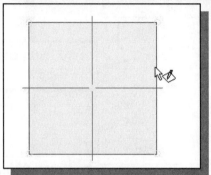

2. In the *Status Bar* area, the message: "*Select face, workplane, sketch or sketch geometry.*" is displayed. Select the *XY Plane* by clicking on any edges of the plane inside the graphics window, as shown.

3. Select the **Project Geometry** command in the *2D Sketch Panel*. The Project Geometry command allows us to project existing features onto the active sketching plane. Left-click once on the icon to activate the Project Geometry command.

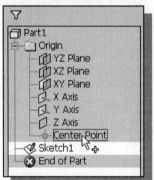

4. In the *Status Bar* area, the message: "*Select edge, vertex, work geometry or sketch geometry to project.*" is displayed. Inside the *Browser* window, select the *Center Point*, *X-axis* and *Y-axis* to project these entities onto the sketching plane.

5. Select the **Line** option in the *2D Sketch Panel*. A *Help-tip box* appears next to the cursor and a brief description of the command is displayed at the bottom of the drawing screen: "*Creates Straight line segments and tangent arcs.*"

6. Create a closed-region sketch with the starting point aligned to the projected Y-axis as shown below. (Note that the *Pulley* design is symmetrical about a horizontal axis as well as a vertical axis, which allows us to simplify the 2D sketch as shown below.)

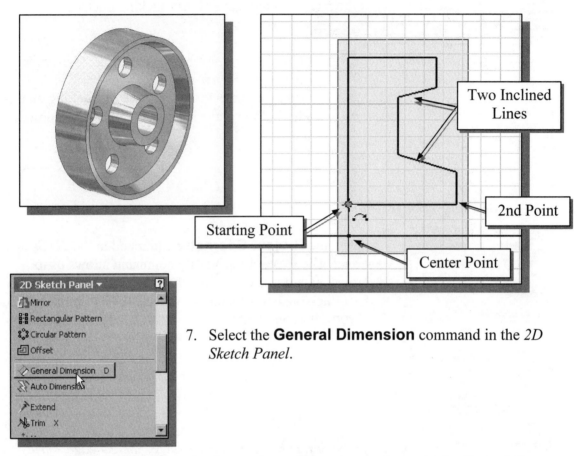

Two Inclined Lines

2nd Point

Starting Point

Center Point

7. Select the **General Dimension** command in the *2D Sketch Panel*.

8. Pick the **X-axis** as the first entity to dimension as shown in the figure below.

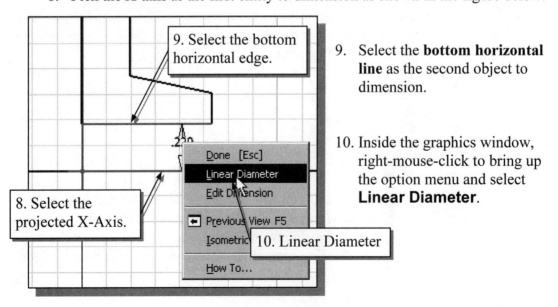

9. Select the bottom horizontal edge.

8. Select the projected X-Axis.

Done [Esc]

Linear Diameter

Edit Dimension

Previous View F5

Isometric

How To...

10. Linear Diameter

9. Select the **bottom horizontal line** as the second object to dimension.

10. Inside the graphics window, right-mouse-click to bring up the option menu and select **Linear Diameter**.

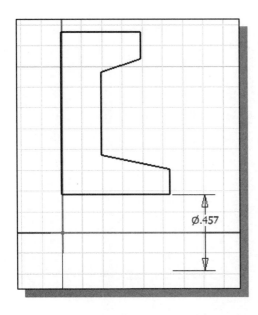

11. Place the dimension text to the right side of the sketch.

- **To create a dimension that will account for the symmetrical nature of the design: pick the axis of symmetry, pick the entity, select Linear Diameter in the option menu, and then place the dimension.**

12. Pick the projected **X-axis** as the first entity to dimension as shown in the figure below.

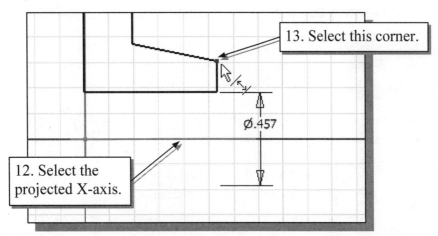

13. Select this corner.

12. Select the projected X-axis.

13. Select the **corner point** as the second object to dimension as shown in the above figure.

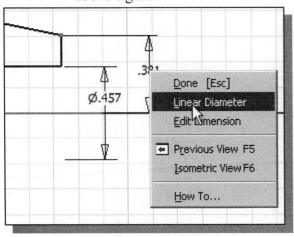

14. Inside the graphics window, right-mouse-click to bring up the option menu and select **Linear Diameter**.

15. Place the dimension text toward the right side of the sketch.

16. On your own, create and adjust the vertical size/location dimensions as shown below. (Hint: Modify the larger dimensions first.)

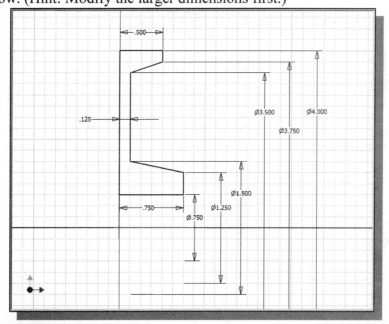

17. Inside the graphics window, click once with the right-mouse-button to display the option menu. Select **Finish Sketch** in the popup menu to end the Sketch option.

➤ On your own, use the **3D-Rotate** command to confirm the completed sketch and dimensions are on a 2D plane.

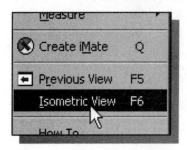

18. Inside the graphics window, right-mouse-click once to bring up the option menu.

19. Select **Isometric View** in the option list to adjust the display of the 2D sketch on the screen.

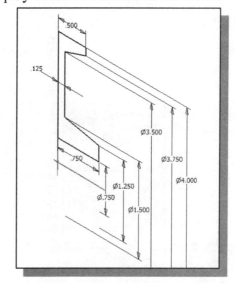

Creating the Revolved Feature

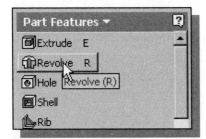

1. In the *Part Features* panel (the panel that is located to the left side of the graphics window), select the **Revolve** command by releasing the left-mouse-button on the icon.

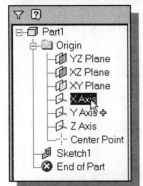

2. In the *Revolve* dialog box, the Axis button is activated indicating *Autodesk Inventor* expects us to select the revolution axis for the revolved feature. Select *X-Axis* as the axis of rotation in the *Browser* window as shown.

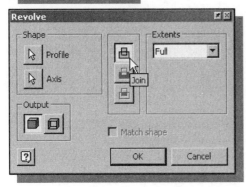

3. In the *Revolve* dialog box, set the termination *Extents* option to **Full** as shown.

4. Click on the **OK** button to accept the settings and create the revolved feature.

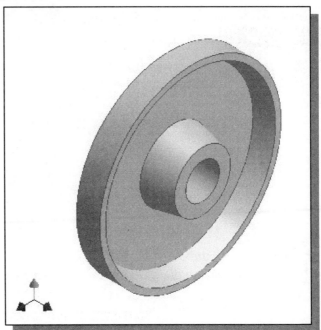

Mirroring Features

- In *Autodesk Inventor*, features can be mirrored to create and maintain complex symmetrical features. We can mirror a feature about a work plane or a specified surface. We can create a mirrored feature while maintaining the original parametric definitions, which can be quite useful in creating symmetrical features. For example, we can create one quadrant of a feature, then mirror it twice to create a solid with four identical quadrants.

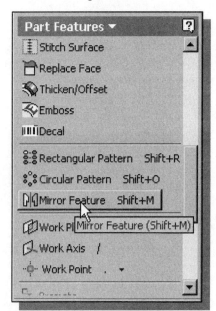

1. In the *Part Features* panel, select the **Mirror Feature** command by releasing the left-mouse-button on the icon.

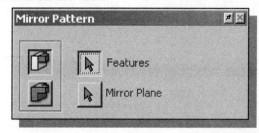

2. In the *Mirror Pattern* dialog box, the **Features** button is activated. *Autodesk Inventor* expects us to select features to be mirrored. In the prompt area, the message *"Select feature to pattern"* is displayed. Select any edge of the 3D base feature.

3. Inside the graphics window, right-mouse-click to bring up the option menu.

4. Select **Continue** in the option list to proceede with the Mirror Feature command.

5. In the *Mirror Pattern* dialog box, the **Mirror Plane** button is activated. *Autodesk Inventor* expects us to select a planar surface about which to mirror. In the prompt area, the message *"Select plane to mirror about"* is displayed.

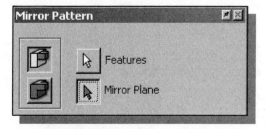

6. On your own, use the **3D-Rotate** function key [**F4**] to dynamically rotate the solid model so that we are viewing the back surface as shown on the next page.

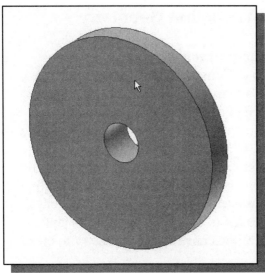

7. Select the surface as shown below as the planar surface about which to mirror.

8. Click on the **OK** button to accept the settings and create a mirrored feature.

9. On your own, use the 3D-Rotate function key [**F4**] to dynamically rotate the solid model and view the resulting solid.

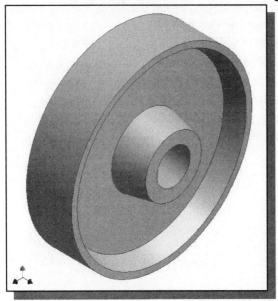

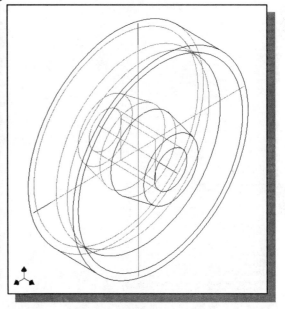

➢ Now is a good time to save the model (Quick key: [**Ctrl**] + [**S**]). It is a good habit to save your model periodically, just in case something might go wrong while you are working on it. You should also save the model after you have completed any major constructions.

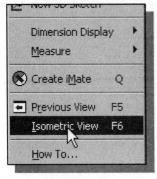

10. Inside the graphics window, right-mouse-click to bring up the option menu.

11. Select **Isometric View** in the option list to adjust the display of the 2D sketch on the screen.

Creating a Pattern Leader Using Construction Geometry

- In *Autodesk Inventor*, we can also use **construction geometry** to help define, constrain, and dimension the required geometry. **Construction geometry** can be lines, arcs, and circles that are used to line up or define other geometry, but are not themselves used as the shape geometry of the model. When profiling the rough sketch, *Autodesk Inventor* will separate the construction geometry from the other entities and treat them as construction entities. Construction geometry can be dimensioned and constrained just like any other profile geometry. When the profile is turned into a 3D feature, the construction geometry remains in the sketch definition but does not show in the 3D model. Using construction geometry in profiles may mean fewer constraints and dimensions are needed to control the size and shape of geometric sketches. We will illustrate the use of the construction geometry to create a cut feature.

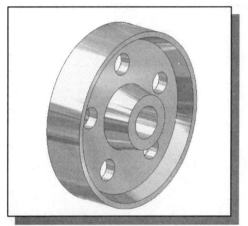

- The *Pulley* design requires the placement of five identical holes on the base solid. Instead of creating the five holes one at a time, we can simplify the creation of these holes by using the **Pattern** command, which allows us to create duplicate features. Prior to using the Pattern command, we will first create a *pattern leader*, which is a regular extruded feature.

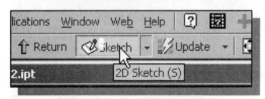

1. In the *Standard* toolbar select the **Sketch** command by left-clicking once on the icon.

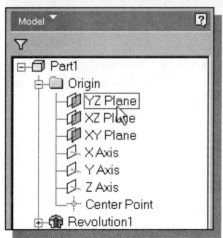

2. In the *Status Bar* area, the message: "*Select face, work plane, sketch or sketch geometry.*" is displayed. Pick the *YZ Plane*, inside the *Browser* window, as shown.

3. Select the **Project Geometry** command in the *2D Sketch Panel*. The **Project Geometry** command allows us to project existing geometry to the active sketching plane. Left-click once on the icon to activate the Project Geometry command.

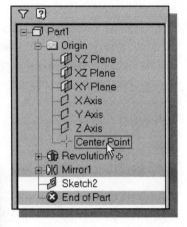

4. In the *Status Bar* area, the message: "*Select edge, vertex, work geometry or sketch geometry to project.*" is displayed. Inside the *Browser* window, select the **Center Point**, **Y-axis** and **Z-axis** to project these entities onto the sketching plane.

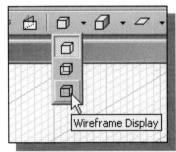

5. On your own, set the display to *wireframe* by clicking on the **Wireframe Display** icon as shown.

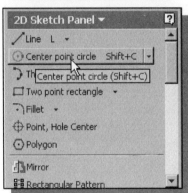

6. Select the **Center point circle** command by clicking once with the left-mouse-button on the icon in the *2D Sketch Panel*.

7. Create a circle of arbitrary size as shown below.

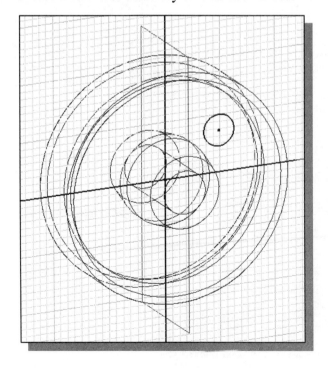

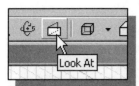

8. In the *Standard* toolbar area, click on the **Look At** button.

9. Select the circle we just created to orient the display of the sketching plane on the screen.

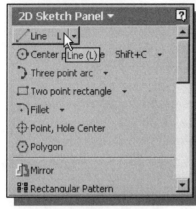

10. Select the **Line** command in the *2D Sketch Panel*. A brief description of the command is displayed at the bottom of the drawing screen: "*Creates Straight line segments and tangent arcs.*"

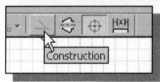

11. Set the *Style* option to **Construction** as shown.

12. Create a *construction line* by connecting from the center of the circle we just created to the projected center point (***origin***) at the center of the 3D model as shown below.

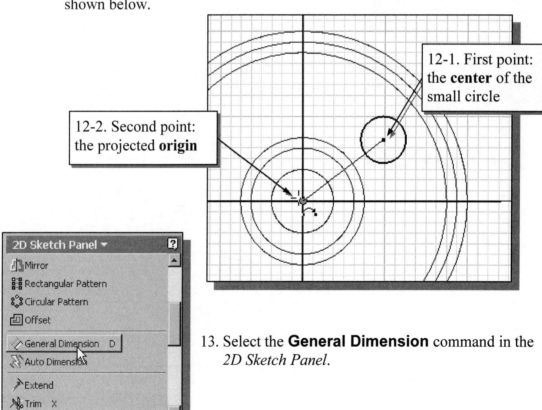

12-1. First point: the **center** of the small circle

12-2. Second point: the projected **origin**

13. Select the **General Dimension** command in the *2D Sketch Panel*.

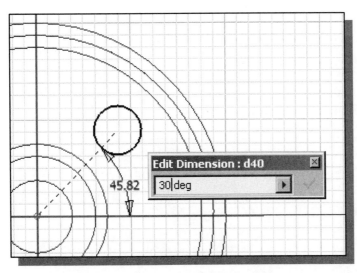

14. Pick the *X-axis* as the first entity to dimension as shown in the figure.

15. Select the **construction line** as the second object to dimension.

16. Place the dimension text to the right of the model as shown.

17. On your own, set the angle dimension to **30** as shown in the above figure.

➤ Note that the location of the small circle is adjusted as the location of the construction line is adjusted by the *angle dimension* we created.

18. Select the **Center point circle** command by clicking once with the left-mouse-button on the icon in the *2D Sketch Panel*.

19. Confirm the *Style* option is still set to **Construction**.

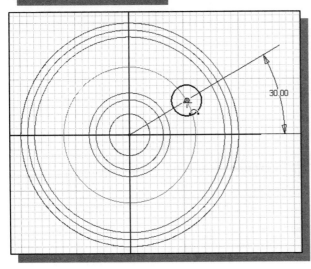

20. Create a *construction circle* by placing the center at the projected center point (*origin*).

21. Pick the center of the small circle to set the size of the construction circle as shown

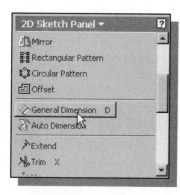

22. Select the **General Dimension** command in the *2D Sketch Panel*.

23. On your own, create the two diameter dimensions, **.5** and **2.5**, as shown below.

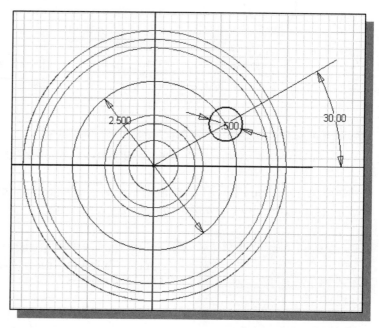

24. Inside the graphics window, click once with the right-mouse-button to display the option menu. Select **Finish Sketch** in the popup menu to end the Sketch option.

25. In the *Part Features* panel (the panel that is located to the left side of the graphics window), select the **Extrude** command by releasing the left-mouse-button on the icon.

26. Pick the inside region of the circle to set up the profile of the extrusion.

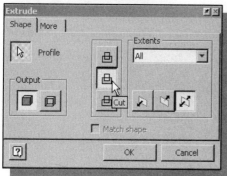

27. Inside the *Extrude* dialog box, select the **Cut** operation, both directions and set the *Extents* to **All** as shown.

28. Click on the **OK** button to accept the settings and create the cut feature.

29. On your own, adjust the angle dimension applied to the construction line of the cut feature to **90** and observe the effect of the adjustment.

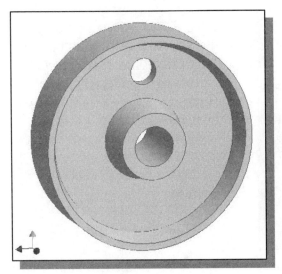 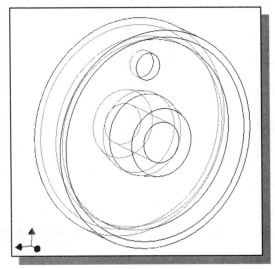

Circular Pattern

- In *Autodesk Inventor*, existing features can be easily duplicated. The **Pattern** command allows us to create both rectangular and polar arrays of features. The patterned features are parametrically linked to the original feature; any modifications to the original feature are also reflected in the arrayed features.

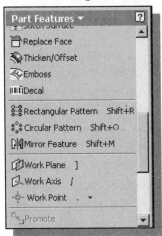

1. In the *Part Features* panel, select the **Circular Pattern** command by left-clicking once on the icon.

2. The message "*Select Feature to be arrayed:*" is displayed in the command prompt window. Select the **circular cut feature** when it is highlighted as shown.

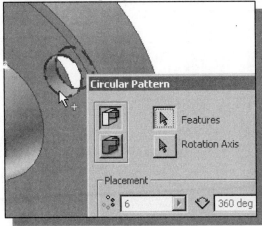

3. Inside the graphics window, right-mouse-click to bring up the option menu.

4. Select **Continue** in the option list to proceed with the Circular Pattern command.

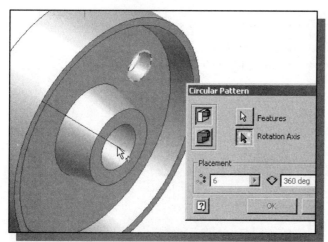

5. In the *Circular Pattern* dialog box, the **Rotation Axis** button is activated. *Autodesk Inventor* expects us to select an axis to pattern about. In the prompt area, the message "*Define the axis of revolution*" is displayed. Select the *X-Axis* in the *Browser* window or in the graphics window.

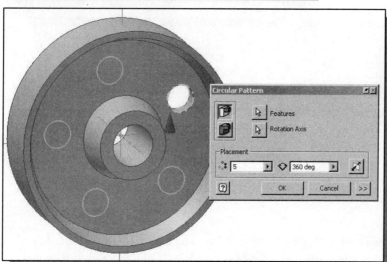

6. In the *Circular Pattern* dialog box, enter **5** in the *Count* box and **360** in the *Angle* box as shown. (Note the different options available for the Circular Pattern command.)

7. Click on the **OK** button to accept the settings and create the *circular pattern*.

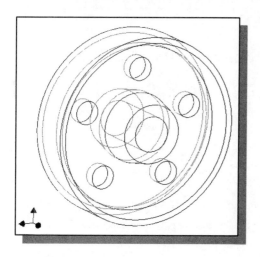

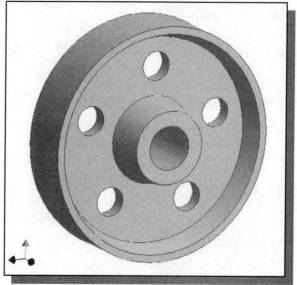

Examining the Design Parameters

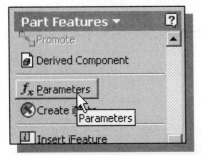

1. In the *Part Features* panel select the **Parameters** command by left-clicking once on the icon. The *Parameters* popup window appears.

2. Note the two dimensions used to create the circular pattern are also listed as shown in the figure below.

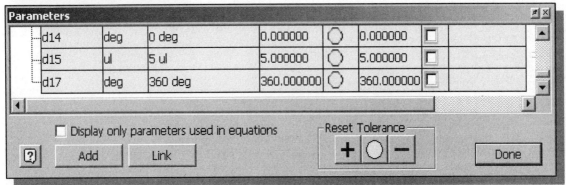

d14	deg	0 deg	0.000000	◯	0.000000	☐	
d15	ul	5 ul	5.000000	◯	5.000000	☐	
d17	deg	360 deg	360.000000	◯	360.000000	☐	

☐ Display only parameters used in equations

Reset Tolerance

3. Click on the **Done** button to accept the settings.

4. Select **Save** in the *Standard* toolbar; we can also use the "**Ctrl-S**" combination (press down the [Ctrl] key and hit the [S] key once) to save the part as *Pulley*.

Drawing Mode – Defining a New Border and Title Block

1. Click on the **drop-down arrow** next to the **New File** button in the *Standard* toolbar area to display the available new file options.

2. Select **Drawing** from the drop-down list.

➢ Note that a new graphics window appears on the screen. We can switch between the solid model and the drawing by clicking the corresponding graphics window.

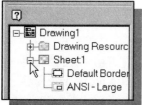

3. Inside the drawing *Browser* window, click on the [**+**] symbol in front of **Sheet1** to display the settings.

4. On your own, **delete** the *Default Border* and the *ANSI title block*. (Hint: right-mouse-click on the names to bring up the option menu.)

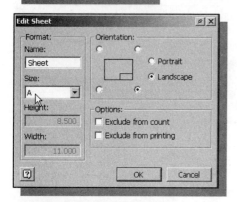

5. Inside the drawing *Browser* window, right-mouse-click on **Sheet1** to bring up the option menu.

6. Select **Edit Sheet** in the option list.

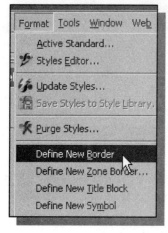

7. In the *Edit Sheet* dialog box, set the sheet size to *A-size (8.5 × 11)*.

8. Confirm the page *Orientation* is set to **Landscape** and click on the **OK** button to accept the settings.

9. In the pull-down menu, select
   Format → Define New Border

10. On your own, create a rectangle (**7.75 × 10.25**) using the **Two Point Rectangle** command and the **General Dimension** command.

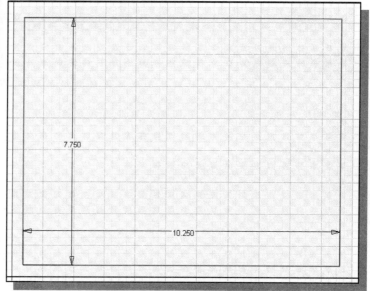

11. Inside the graphics window, right-mouse-click to bring up the option menu.

12. Select **Save Border** in the option list.

13. In the *Border* dialog box, enter *A-Size* as the new border name.

14. Click on the **Save** button to end the Define New Border command.

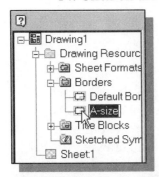

15. Inside the *Browser* window, expand the ***Drawing Resources*** list by clicking on the [**+**] symbol.

16. Double-click on the *A-size* border, the border we just created, to place the border in the current drawing. Note that none of the dimensions used to construct the border is displayed.

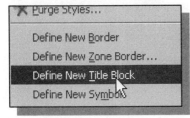

17. In the pull-down menu, select **[Format] → [Define New Title Block]**

18. On your own, create a title block using the **Two Point Rectangle**, **Line**, **General Dimension,** and **Text** commands.

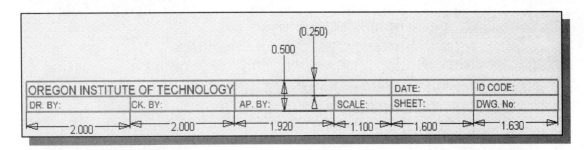

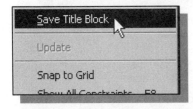

19. Inside the graphics window, right-mouse-click to bring up the option menu.

20. Select **Save Title Block** in the option list.

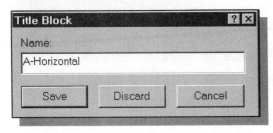

21. In the *Title Block* dialog box, enter **A-Horizontal** as the new title block name.

22. Click on the **Save** button to end the Define New Title Block command.

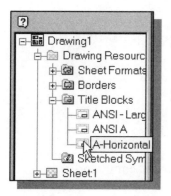

23. Inside the *Browser* window, expand the ***Title Blocks*** list by clicking on the [**+**] symbol.

24. Double-click on the ***A-Horizontal*** title block, the title block we just created, to place it into the current drawing.

Creating a Drawing Template

> In *Autodesk Inventor*, each new drawing is created from a template. During the installation of *Autodesk Inventor*, a default drafting standard was selected which sets the default template used to create drawings. We can use this template or another predefined template, modify one of the predefined templates, or create our own templates to enforce drafting standards and other conventions. Any drawing file can be used as a template; a drawing file becomes a template when it is saved in the *Templates* folder. Once the template is saved, we can create a new drawing file using the new template.

1. Select **Save** in the *Standard* toolbar, we can also use the "**Ctrl-S**" combination (press down the [**Ctrl**] key and hit the [**S**] key once) to save the part.

2. In the *Save As* dialog box, switch to the **Inventor English Templates** directory.

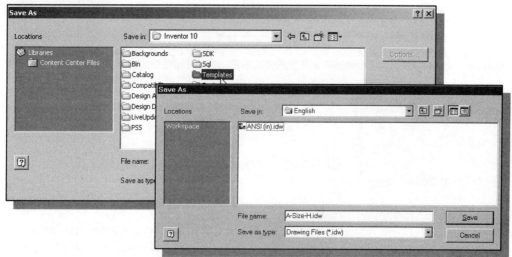

3. Enter ***A-Size-H.idw*** as the template filename.

4. Click on the **Save** button to create a drawing template file.

Automatically Retrieve Dimensions

> ➢ So far, parametric/model dimensions have been retrieved manually. *Autodesk Inventor* also allows us to retrieve dimensions when the views are created.

1. Select **Tools → Application Options** to open the *Options* dialog box.

2. On the **Drawing** tab, switch *ON* the **Retrieve all model dimensions on view placement** option.

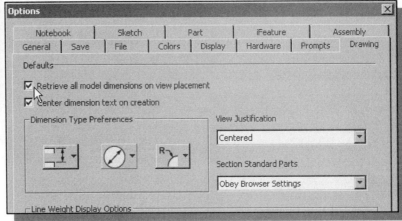

Creating Views

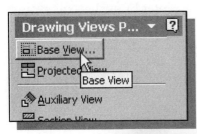

1. Click on **Base View** in the *Drawing Views Panel* to create a base view.

2. In the *Drawing View* dialog box, set the scale to **0.75** and select the *left* view as shown in the figure below. (**DO NOT** click on the **OK** button.)

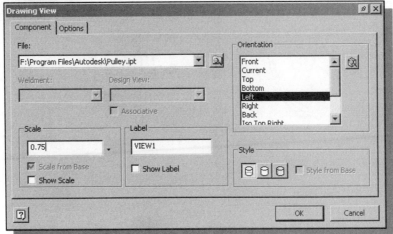

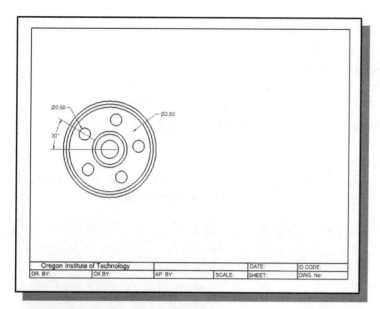

3. Move the cursor inside the graphics window and place the **base view** toward the left side of the border as shown. (If necessary, drag the *Drawing View* dialog box to another location on the screen.) Note that some of the associated dimensions are displayed as the view is created.

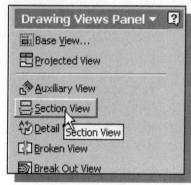

4. Click on the **Section View** button in the *Drawing Views Panel* to create a section view.

5. Click on the **base view** to select the view where the section line is to be created.

6. Inside the graphics window, align the cursor to the center of the base view and create the vertical cutting plane line as shown.

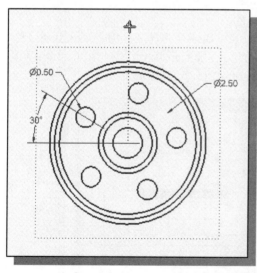

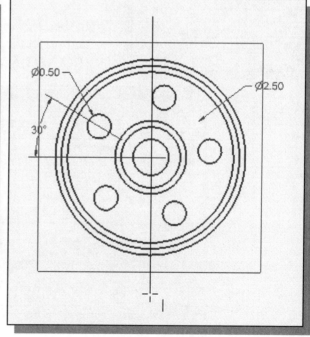

7. Inside the graphics window, right-mouse-click once to bring up the option menu.

8. Select **Continue** to proceed with the Section View command.

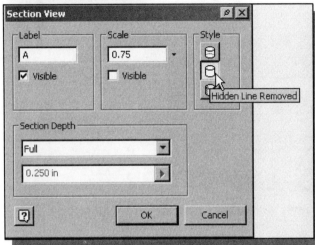

9. Inside the *Section View* dialog box, turn *OFF* the **Show Scale** option.

10. Set the *Display Style* to **Hidden Line Removed**, as shown in the above figure. (**DO NOT** click on the **OK** button.)

11. Next, *Autodesk Inventor* expects us to place the projected section. Select a location that is toward the right side of the base view as shown in the figure.

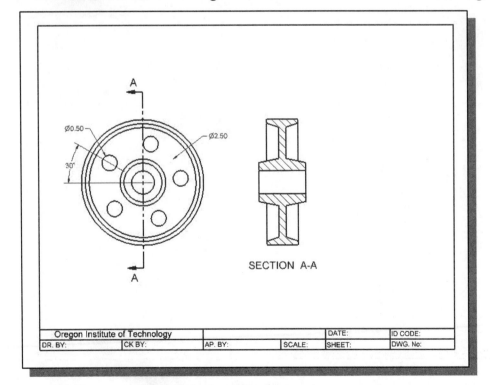

➢ Note that, since *Autodesk Inventor R8*, the parametric/model dimensions are no longer automatically retrieved when the section view is created.

12. On your own, use the **Base View** option and create an **isometric view** (Iso Top Left) of the design and place the view toward the right side of the section view as shown below.

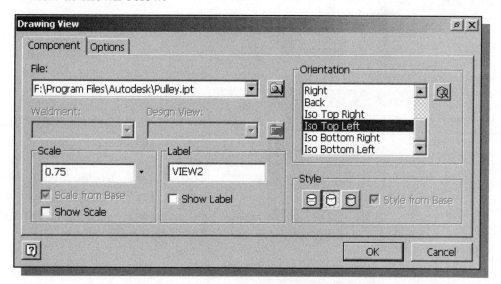

➢ On your own, reposition the views so that they appear as shown in the figure below.

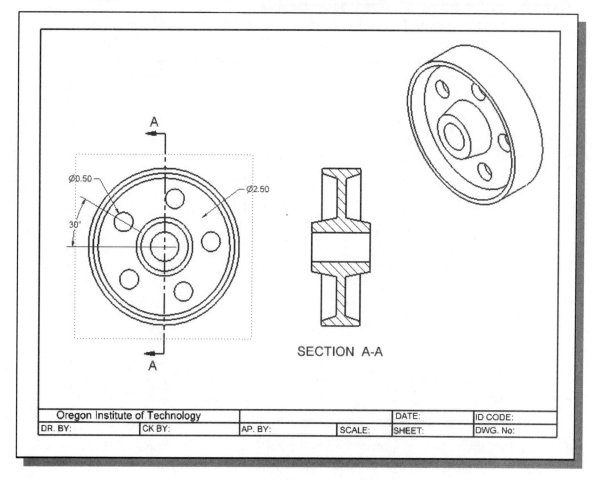

Retrieve Dimensions – Features Option

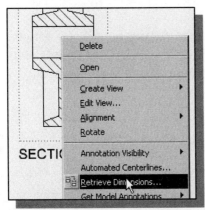

1. Move the cursor on top of the **Section view** of the model and watch for the box around the entire view indicating the view is selectable as shown in the figure.

2. Inside the graphics window, right-mouse-click once to bring up the option menu.

3. Select **Retrieve Dimensions** to display the parametric dimensions used to create the model.

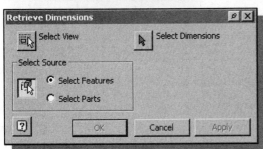

4. In the *Retrieve Dimensions* dialog box, set the *Select Source* option to **Select Features** as shown.

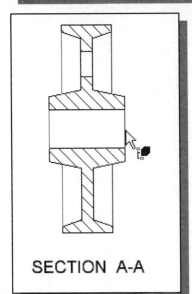

5. Move the cursor to the **Section view** and select one of the right edges of the main body as shown. (Note that the parametric dimensions are applied to the right side of the main body.)

➢ The dimensions used to create the revolved feature are now displayed in the section view.

6. Click on the **Select Dimensions** button to activate the command.

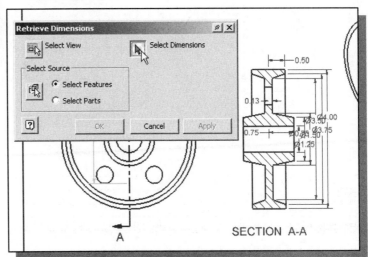

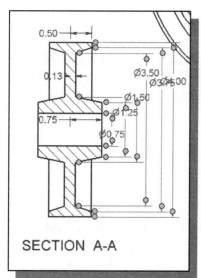

SECTION A-A

7. Select all the displayed **diameter dimensions** as shown in the figure.

8. Click on the **Apply** button and notice the selected dimensions changed color indicating they have been retrieved.

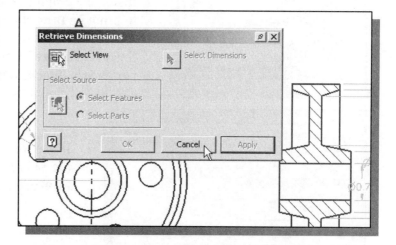

9. Click on the **Cancel** button to end the Retrieve Dimensions command.

10. On your own, create the additional horizontal dimensions and adjust the positions of the dimensions of the section view.

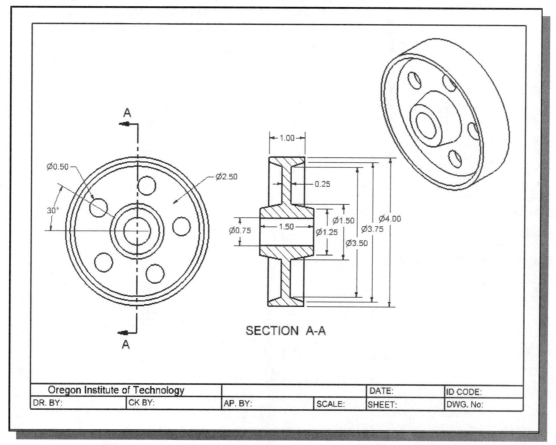

Associative Functionality – A Design Change

- *Autodesk Inventor's associative functionality* allows us to change the design at any level, and the system reflects the changes at all levels automatically. We will illustrate the associative functionality by changing the circular pattern from five holes to six holes.

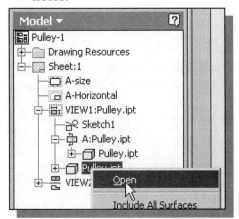

1. Inside the *Model Tree* window, below the VIEW1 list, right-click on the *Pulley.ipt* part name to bring up the option menu.

2. Select **Open** in the popup menu to switch to the associated solid model.

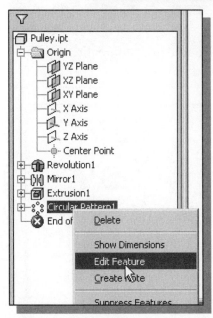

3. Inside the *Model Tree* window, right-click on the *CircularPattern1* feature to bring up the option menu.

4. Select **Edit Feature** in the popup menu to bring up the associated feature option.

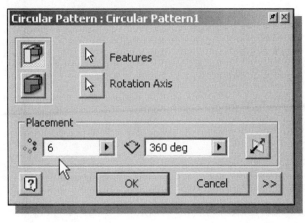

5. In the *Circular Pattern* dialog box, change the number to **6** as shown.

6. Click on the **OK** button to accept the setting.

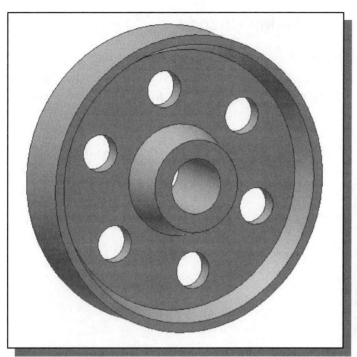

❖ The solid model is updated showing the 6 equally spaced holes as shown.

7. Switch back to the *Pulley* drawing and notice the drawing is also updated.

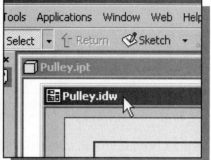

❖ Notice, in the *Pulley* drawing, the circular pattern is also updated automatically in all views.

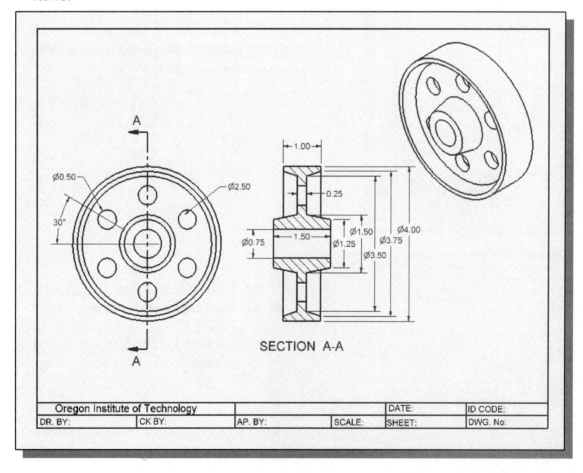

Adding Centerlines to the Pattern feature

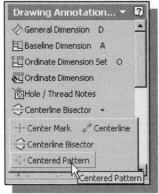

1. Click on the **drop-down arrow** next to the **Center Mark** button in the *Drawing Annotation* window to display the available options.

2. Select **Centered Pattern** from the option list.

   ➤ The **Centered Pattern** option allows us to add centerlines to a patterned feature.

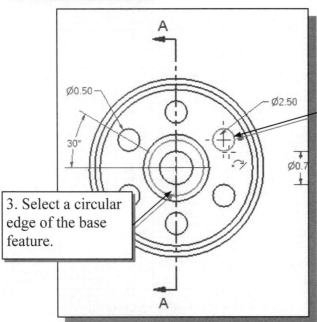

3. Select a circular edge of the base feature.

4. Select any circular edge of the patterned feature.

3. Inside the graphics window, click on any circular edge of the base feature.

4. Select any circular edge that is part of the patterned feature.

5. Continue to select the circular edges of the patterned features, in a counterclockwise manner, until all patterned items are selected. (Select the first circle as the ending item.)

6. Inside the graphics window, right-mouse-click once to bring up the option menu.

7. Select **Create** to create the centerlines around the selected items.

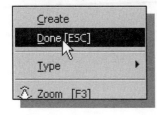

8. Inside the graphics window, right-mouse-click once to bring up the option menu.

9. Select **Done** to end the Centered Pattern command.

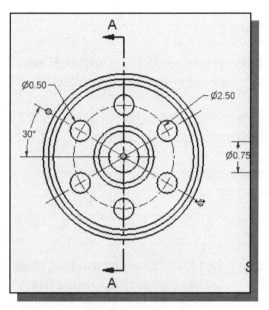

10. On your own, extend the segments of the centerlines so that they pass through the center of the base feature.

Completing the Drawing

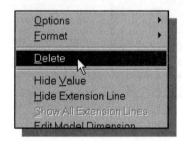

1. On your own, use **Delete** from the option menu to delete unwanted dimensions.

2. Use the **General Dimension** command to create additional dimensions as shown below.

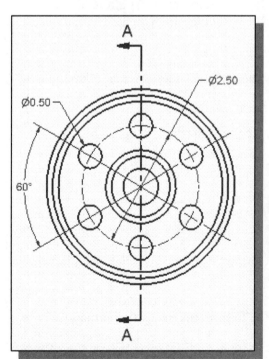

➢ For the circular pattern, confirm with the right-mouse-button that the **Arrowheads Inside** option is switched *ON* and switched *OFF* the **Single Dimension Line** option.

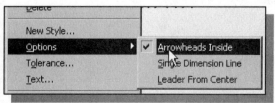

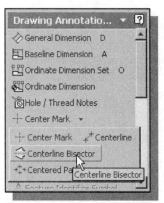

3. Click on the **drop-down arrow** next to the **Center Mark** button in the *Drawing Annotation* window to display the available options.

4. Select **Centerline Bisector** from the option list.

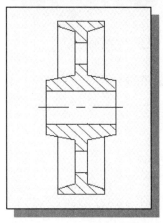

5. Inside the graphics window, click on the two edges of the *section view* as shown in the figure to create a centerline through the view.

6. Repeat the above step and create the centerlines through the patterned holes.

7. On your own, complete the drawing by completing the title block.

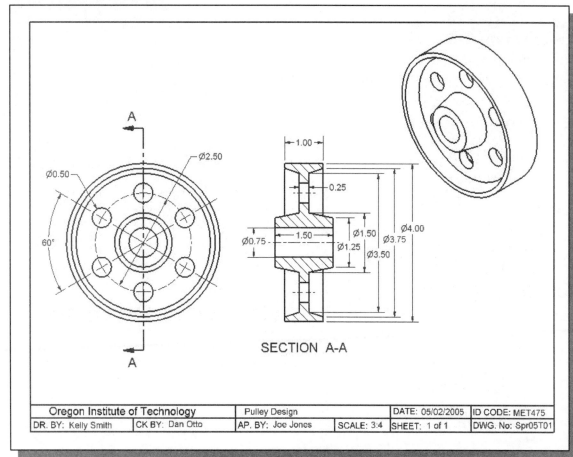

Questions:

1. List the different symmetrical features created in the *Pulley* design.

2. What are the advantages of using a *drawing template*?

3. Describe the steps required in using the **Mirror Feature** command.

4. Why is it important to identify symmetrical features in designs?

5. When and why should we use the **Pattern** option?

6. What is the difference between *Rectangular Pattern* and *Circular Pattern*?

7. How do we create a *Linear Diameter dimension* for a revolved feature?

8. What is the difference of *construction geometry* and *normal geometry*?

9. Identify and describe the following commands:

   (a)

   (b)

   (c)

   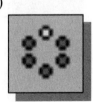

   (d)

Exercises: (All dimensions are in inches.)

1.

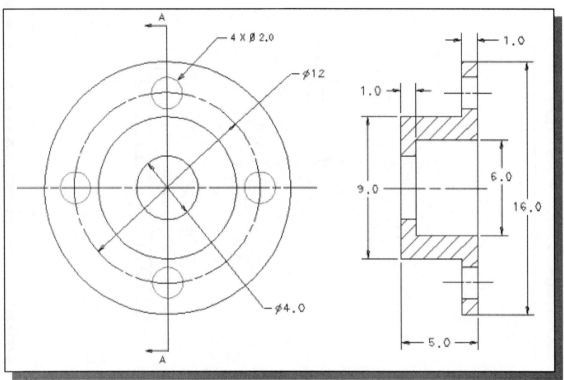

2. Plate thickness: 0.125 inch

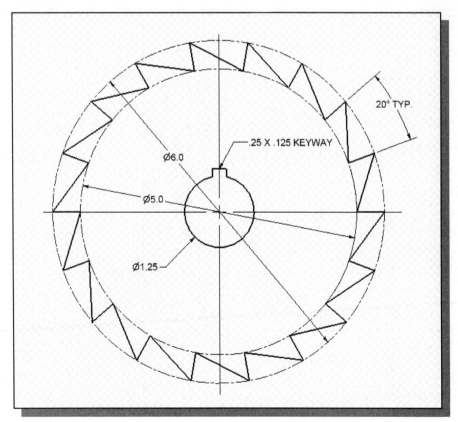

3.

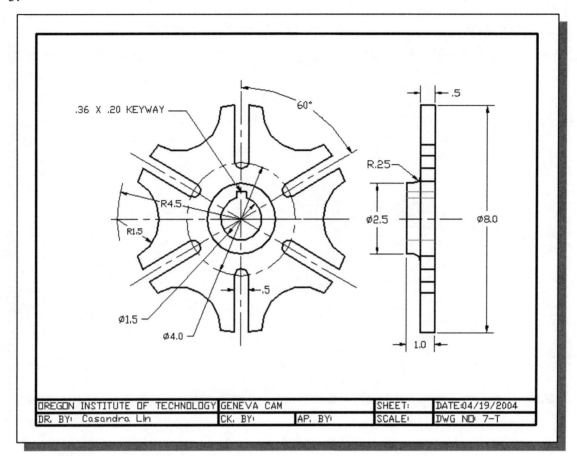

.36 X .20 KEYWAY

60°

R.25

R4.5

Ø2.5

Ø8.0

R1.5

.5

.5

Ø1.5

Ø4.0

1.0

OREGON INSTITUTE OF TECHNOLOGY	GENEVA CAM		SHEET:	DATE:04/19/2004
DR. BY: Casandra Lin	CK. BY:	AP. BY:	SCALE:	DWG NO: 7-T

Chapter 11
Advanced 3D Construction Tools

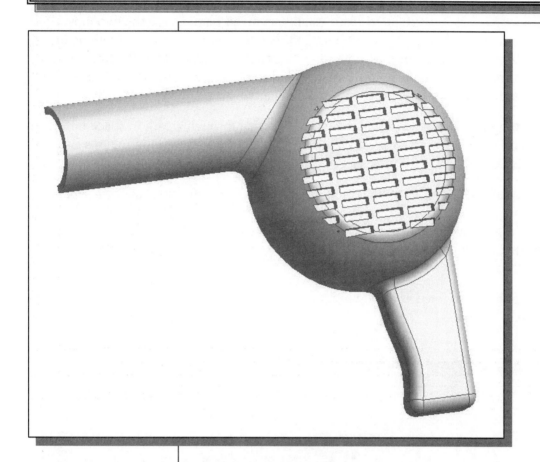

Learning Objectives

♦ **Understand the Concepts Behind the Different 3D Construction Tools**
♦ **Setup Multiple Work Planes**
♦ **Create Swept Features**
♦ **Create Lofted Features**
♦ **Use the Shell Command**
♦ **Create 3D Rounds & Fillets**

Introduction

Autodesk Inventor provides an assortment of three-dimensional construction tools to make the creation of solid models easier and more efficient. As demonstrated in the previous lessons, creating **extruded** features and **revolved** features are the two most common methods used to create 3D models. In this next example, we will examine the procedures for using the **Sweep** command, the **Loft** command, and the **Shell** command, and also for creating **3D rounds** and **fillets** along the edges of a solid model. These types of features are common characteristics of molded parts.

The **Sweep** option is defined as moving a cross-section through a path in space to form a three-dimensional object. To define a sweep in *Autodesk Inventor*, we define two sections: the trajectory and the cross-section.

The **Loft** command allows us to blend multiple profiles with varying shapes on separate planes to create complex shapes. Profiles are usually on parallel planes, but non-perpendicular planes can also be used. We can use as many profiles as we wish but, to avoid twisting the loft shape, we should map points on each profile that align along a straight vector.

The **Shell** option is defined as hollowing out the inside of a solid, leaving a shell of specified wall thickness.

A Thin-Walled Design: *Dryer Housing*

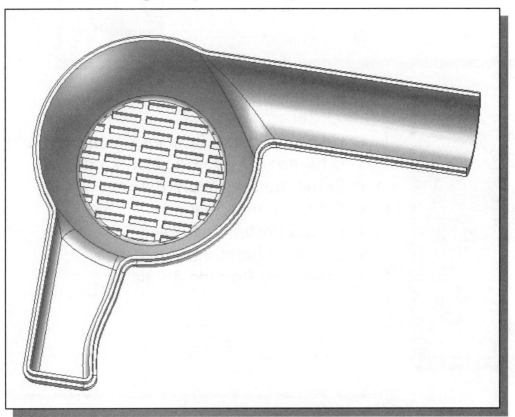

Modeling Strategy

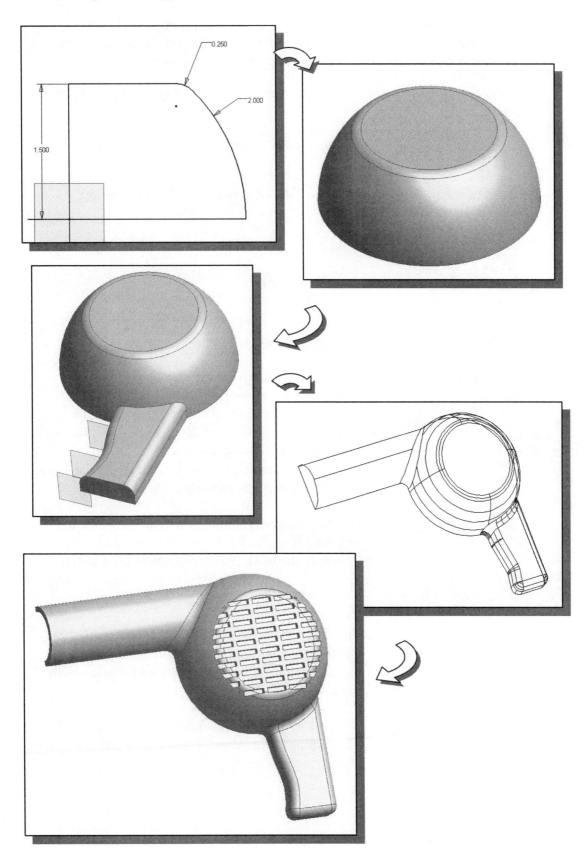

Starting Autodesk Inventor

1. Select the **Autodesk Inventor** option on the *Start* menu or select the **Autodesk Inventor** icon on the desktop to start *Autodesk Inventor*. The *Autodesk Inventor* main window will appear on the screen.

2. Once the program is loaded into the memory, the ***Startup*** dialog box appears at the center of the screen.

3. Select the **New** icon with a single click of the left-mouse-button in the *What to Do* dialog box.

4. Select the **English** tab and in the *New - Choose Template* area, select **Standard(in).ipt**.

5. Pick **OK** in the *Startup* dialog box to accept the selected settings.

Set up the Display of the Sketch Plane

1. In the part *Browser* window, click on the [**+**] symbol in front of the ***Origin*** feature to display more information on the feature.

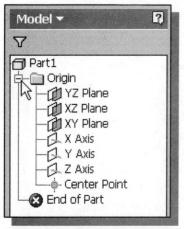

❖ In the *Browser* window, notice a new part name appeared with seven work features established. The seven work features include three *workplanes*, three *work axes*, and a *work point*. By default, the three work planes and work axes are aligned to the **world coordinate system** and the work point is aligned to the *origin* of the **world coordinate system**.

2. Inside the *Browser* window, move the cursor on top of the third work plane, ***XY Plane***. Notice a rectangle, representing the work plane, appears in the graphics window.

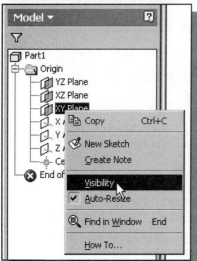

3. Inside the *Browser* window, click once with the right-mouse-button on ***XY Plane*** to display the option menu. Click on **Visibility** to toggle *ON* the display of the plane.

4. On your own, repeat the above step and toggle *ON* the display of the *X* and *Y work axes* and the ***Center Point*** on the screen.

Creating the 2D Sketch for the Base Feature

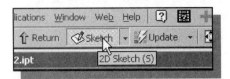

1. In the *Standard* toolbar select the **Sketch** command by left-clicking once on the icon.

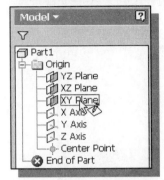

2. In the *Status Bar* area, the message: "*Select face, workplane, sketch or sketch geometry.*" is displayed. Select the ***XY Plane***, by left-clicking once on any edges of the XY Plane in the graphics window or in the *Browser* window as shown.

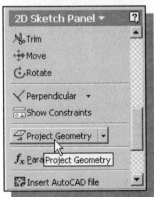

3. Select the **Project Geometry** command in the *2D Sketch Panel*. The Project Geometry command allows us to project existing features to the active sketching plane. Left-click once on the icon to activate the Project Geometry command.

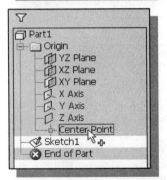

4. In the *Status Bar* area, the message: "*Select edge, vertex, work geometry or sketch geometry to project*" is displayed. Inside the *Browser* window, select the ***Center Point*** to project the center point onto the sketching plane.

5. Select the **Center point circle** command by clicking once with the left-mouse-button on the icon in the *2D Sketch Panel*.

6. Pick the **projected center point** as the center location of the circle.

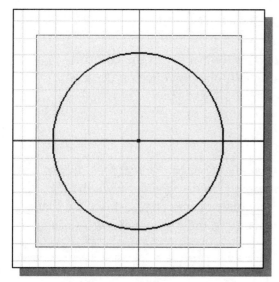

7. Create a **circle** of arbitrary size, with its center aligned to the projected center point as shown.

8. Click on the **Line** icon in the *2D Sketch Panel*.

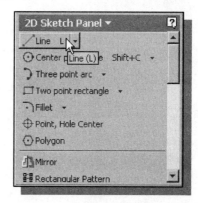

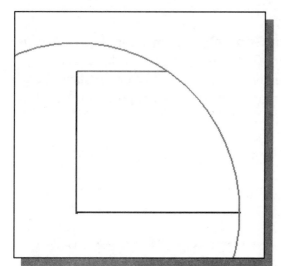

9. Create the three line segments as shown. The lines are either horizontal or vertical with the lower left corner aligned to the projected *Center Point*.

10. Click on the **Trim** command in the *2D Sketch Panel*.

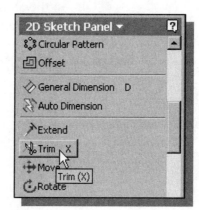

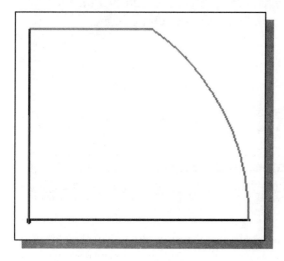

11. Modify the sketch by trimming the circle as shown.

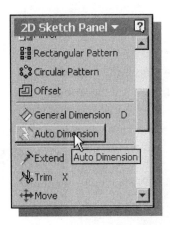

12. On your own, use the **Auto Dimension** command and modify the dimensions as shown in the figure below. (Hint: modify the radius dimension first.)

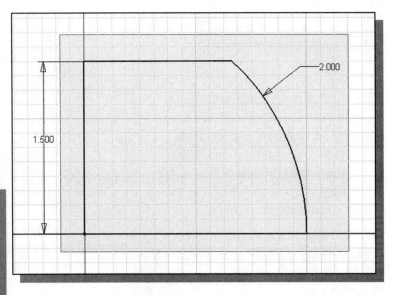

13. Click on the **Fillet** icon in the *2D Sketch Panel*.

14. In the *2D Fillet* dialog box, set the *radius* to **0.25**.

15. Select the top horizontal line and the arc to create a rounded corner as shown.

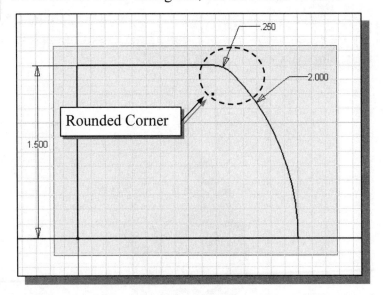

16. Inside the graphics window, click once with the right-mouse-button to display the option menu. Select **Finish Sketch** in the popup menu to end the Sketch option.

Create a Revolved Feature

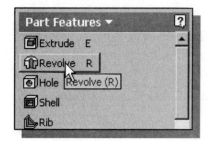

1. In the *Part Features* toolbar (the toolbar that is located to the left side of the graphics window), select the **Revolve** command by left-mouse-clicking the icon.

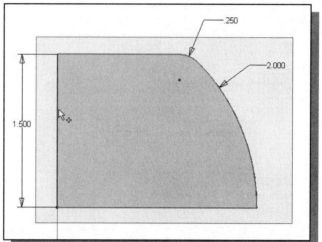

2. In the *Revolve* dialog box, the **Axis** button is activated indicating *Autodesk Inventor* expects us to select the revolution axis for the revolved feature. Select the **vertical edge** of the sketch as the axis of rotation as shown.

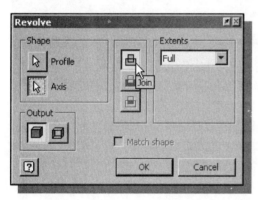

3. In the *Revolve* dialog box, set the termination *Extents* option to **Full** as shown.

4. Click on the **OK** button to accept the settings and create the revolved feature.

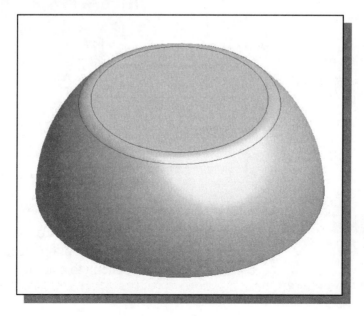

Creating Offset Work Planes

1. Inside the graphics window, right-mouse-click to bring up the option menu.

2. Select **Isometric View** in the option list to adjust the display of the 3D object on the screen.

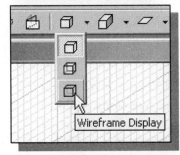

3. Click on the **Wireframe** icon to set the display mode.

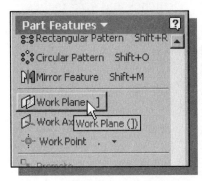

4. In the *Part Features* toolbar, select the **Work Plane** command by left-mouse-clicking on the icon.

5. In the *Status Bar* area, the message: "*Define work plane by highlighting and selecting geometry*" is displayed. *Autodesk Inventor* expects us to select any existing geometry, which will be used as a reference to create the new work plane.

6. Select the *XY Plane* in the graphics window, or in the *Browser* window, as the reference plane. Left-click once to set the *XY Plane* as the reference of a new work plane.

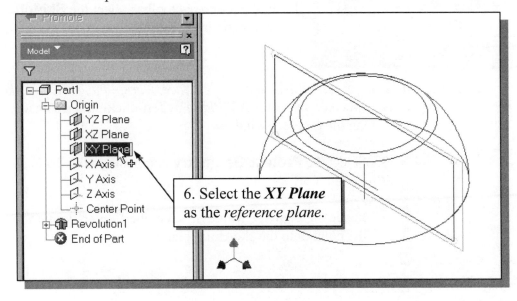

6. Select the *XY Plane* as the *reference plane*.

7. Inside the graphics window, use the left-mouse-button, and drag the highlighted reference plane toward the lower-left corner of the graphics window.

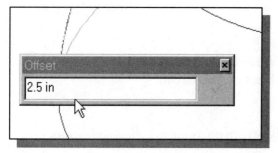

8. In the **Offset** popup window, enter **2.5** as the offset distance for the new work plane.

9. Click on the **check mark** button to accept the setting.

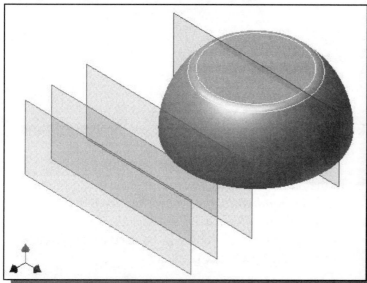

10. On your own, repeat the above steps and create two additional work planes that are **3.5** inches and **4.25** inches away from the *XY Plane*.

Creating 2D Sketches on the Work Planes

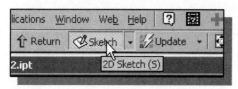

1. In the *Standard* toolbar select the **Sketch** command by left-clicking once on the icon.

2. In the *Status Bar* area, the message: "*Select face, workplane, sketch or sketch geometry*" is displayed. Select the *XY Plane* by left-clicking once on any edge of the *XY Plane* in the graphics window.

3. Select the **Project Geometry** command in the *2D Sketch Panel*.

4. Inside the *Browser* window, select the *Center Point* to project the point onto the sketching plane.

5. Select the **Two point rectangle** command by clicking once with the left-mouse-button on the icon in the *2D Sketch Panel*.

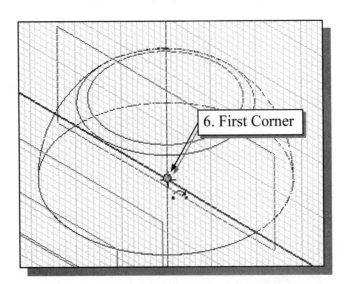

6. Click on the projected *Center Point* to align the first corner of the rectangle.

7. Create a rectangle of arbitrary size by selecting a location that is toward the right side of the graphics window as shown.

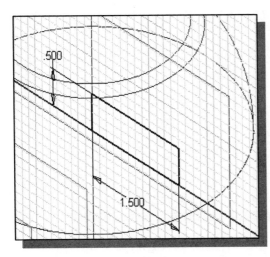

8. On your own, create and modify the two dimensions as shown in the figure.

9. On your own, create two rounded corners (**radius 0.25**) as shown.

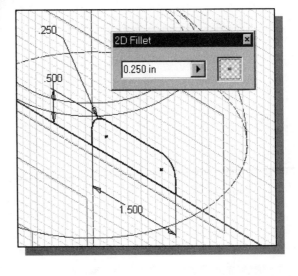

10. Inside the graphics window, click once with the right-mouse-button to display the option menu. Select **Finish Sketch** in the popup menu to end the **Sketch** option.

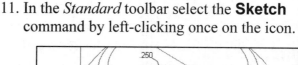

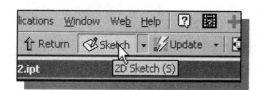

11. In the *Standard* toolbar select the **Sketch** command by left-clicking once on the icon.

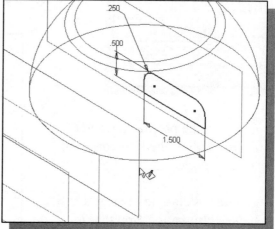

12. In the *Status Bar* area, the message: "*Select face, workplane, sketch or sketch geometry*" is displayed. Select **WorkPlane1**, by left-clicking once on any edge of the first offset work plane in the graphics window as shown.

13. Select the **Project Geometry** command in the *2D Sketch Panel*.

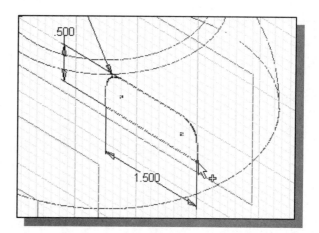

14. Select the lower right corner of the 2D sketch we just created, as shown in the figure.

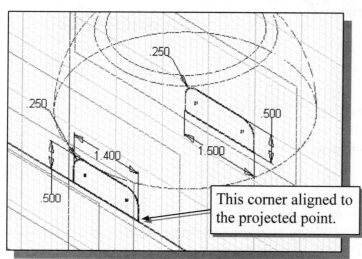

This corner aligned to the projected point.

15. On your own, create and modify the 2D sketch as shown in the figure. (The lower right corner of the sketch is aligned to the lower right corner of the previous sketch.)

16. On your own, repeat the above steps and create two additional sketches on **WorkPlane2** and **WorkPlane3** as shown in the figure below. Note that the lower right corners of the four sketches are aligned in the Z-direction through the use of the projected center point.

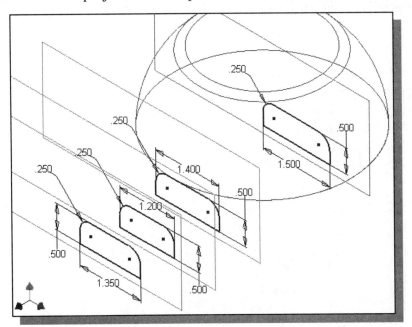

Creating a Lofted Feature

➢ The **Loft** option allows us to blend multiple profiles with varying shapes on separate planes to create complex shapes. Profiles are usually on parallel planes, but any non-perpendicular planes can also be used. We can use as many profiles as we wish, but to avoid twisting the loft shape, we should map points on each profile that align along a straight vector.

1. In the *Part Features* toolbar, select the **Loft** command by left-mouse-clicking on the icon.

2. In the *Loft* dialog box, the **Curves** option is activated. *Autodesk Inventor* expects us to select any number of existing profiles, which will be used to create the lofted feature.

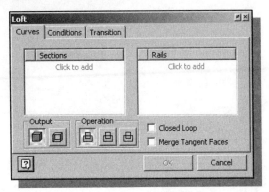

3. Click inside the Sections box and pick the four sketched sections, in the order that they were created, by clicking near the lower right corners of the sketches. Three arrows are displayed showing the blending direction of the sketches. Note that, to avoid twisting the loft shape, we are mapping the same corner points on the sketches to align along a straight vector.

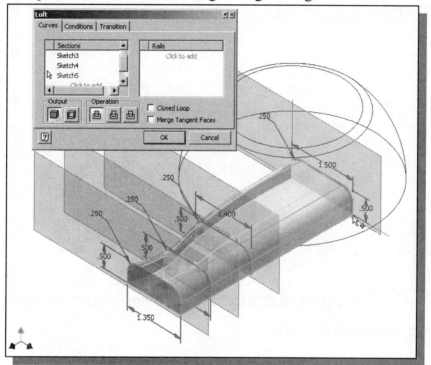

4. Click on the **OK** button to accept the settings and create the lofted feature.

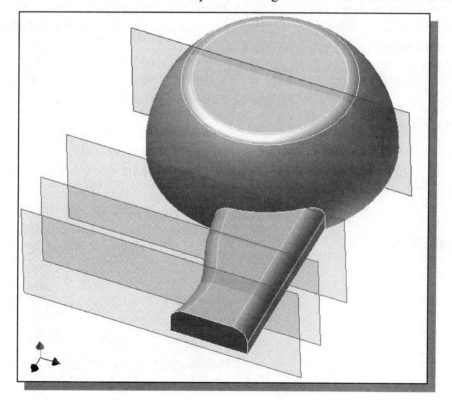

Creating an Extruded Feature

1. In the *Standard* toolbar select the **Sketch** command by left-clicking once on the icon.

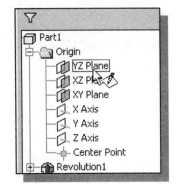

2. In the *Status Bar* area, the message: "*Select face, workplane, sketch or sketch geometry.*" is displayed. Select the ***YZ Plane***, by left-clicking once on the *YZ Plane,* in the *Browser* window.

3. Select the **Project Geometry** command in the *2D Sketch Panel*.

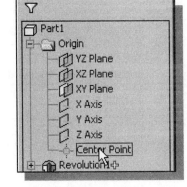

4. In the *Status Bar* area, the message: "*Select edge, vertex, work geometry or sketch geometry to project.*" is displayed. Inside the *Browser* window, select the ***Center Point*** to project the center point onto the sketching plane.

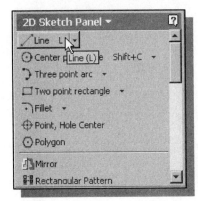

5. Click on the **Line** icon in the *2D Sketch Panel.*

6. Create a line that is parallel (nearly aligned) to the horizontal axis as shown.

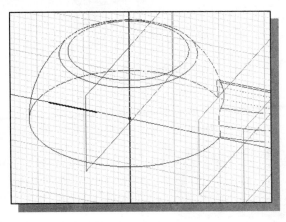

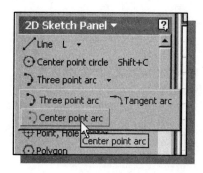

7. Select the **Center point arc** option in the *2D Sketch Panel* as shown.

8. On your own create an arc aligned to the mid-point and endpoints of the previously created line as shown in the figure below.

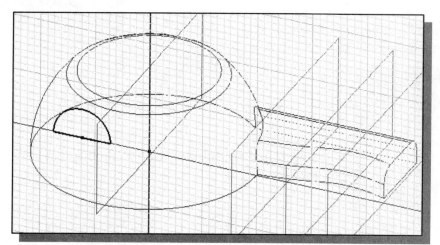

9. On your own, create and modify the three dimensions as shown in the figure below. (Hint: Use the projected center point as the reference point.)

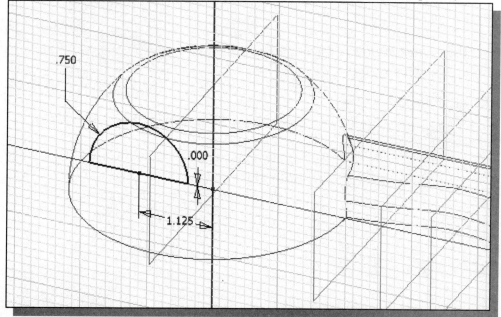

10. Inside the graphics window, click once with the right-mouse-button to display the option menu. Select **Finish Sketch** in the popup menu to end the Sketch option.

Completing the Extruded Feature

1. In the *Part Features* toolbar, select the **Extrude** command by left-clicking once on the icon.

2. Select the inside region of the 2D sketch to create a profile as shown.

3. In the *Extrude* popup window, enter **5.5** as the extrusion distance.

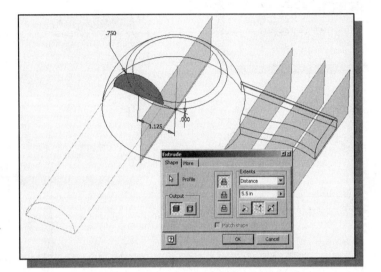

4. In the *Extrude* dialog box, set the extrusion direction as shown and confirm the operation is set to **Join,** and then click on the **OK** button to accept the settings to create the feature.

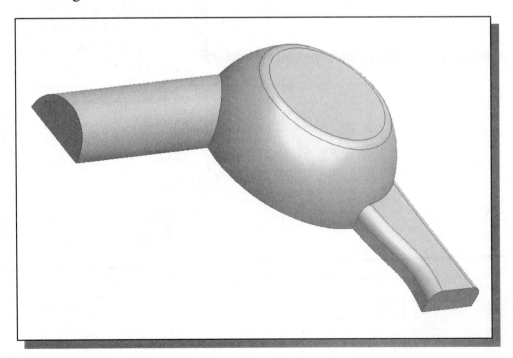

Create 3D Rounds and Fillets

1. In the *Part Features* toolbar, select the **Fillet** command by left-clicking once on the icon.

2. In the *Fillet* dialog box, set the *Radius* option to a radius of **0.2** as shown below. (**DO NOT** click on the **OK** button.)

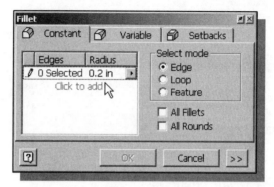

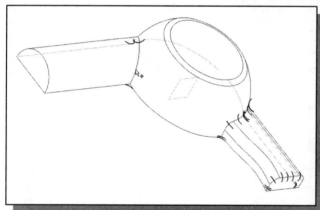

3. Click inside the *Edges* box (*0 selected*) to begin selecting edges.

4. Click on the three edges as shown.

5. Click on the **OK** button to accept the settings and create the 3D rounds and fillets.

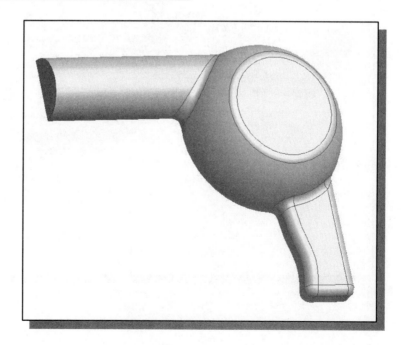

Creating a Shell Feature

- The **Shell** command can be used to hollow out the inside of a solid, leaving a shell of specified wall thickness.

1. In the *Part Features* toolbar, select the **Shell** command by left-clicking once on the icon.

2. On your own, use the **3D Rotation** quickkey [**F4**] to display the back faces of the model as shown below.

3. In the *Shell* dialog box, the **Remove Faces** option is activated. Select the two faces as shown below.

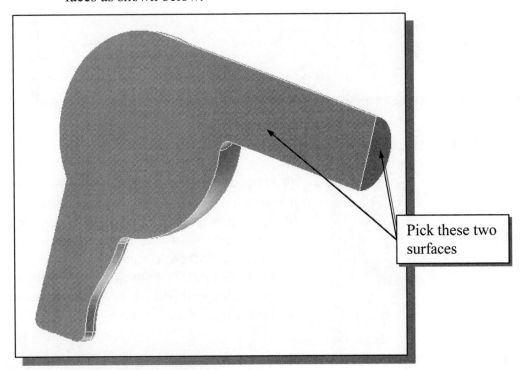

Pick these two surfaces

4. In the *Shell* dialog box, set the option to **Inside** with a value of **0.125** as shown.

5. In the *Shell* dialog box, click on the **OK** button to accept the settings and create the shell feature.

Create a Pattern Leader

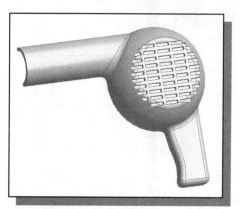

- The *Dryer Housing* design requires the placement of identical holes on the top face of the solid. Instead of creating the holes one at a time, we can simplify the creation of these holes by using the **Pattern** command to create duplicate features. Prior to using the Pattern command, we will first create a *pattern leader*, which is a regular cut feature.

1. Inside the graphics window, right-mouse-click to bring up the option menu.

2. Select **Isometric View** in the option list to adjust the display of the model on the screen.

3. In the *Standard* toolbar select the **Sketch** command by left-clicking once on the icon.

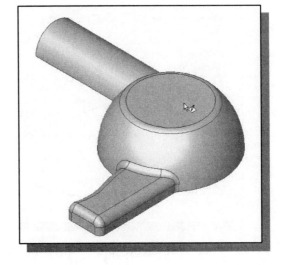

4. In the *Status Bar* area, the message: "*Select face, workplane, sketch or sketch geometry.*" is displayed. Pick the top face of the base feature as shown.

5. Select the **Project Geometry** command in the *2D Sketch Panel*.

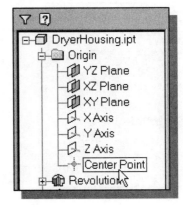

6. In the *Status Bar* area, the message: "*Select edge, vertex, work geometry or sketch geometry to project*" is displayed. Inside the *Browser* window, select **Center Point** to project the point onto the sketching plane.

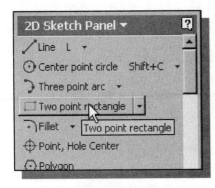

7. Select the **Two point rectangle** command by clicking once with the left-mouse-button on the icon in the *2D Sketch Panel*.

8. Create a rectangle of arbitrary size that is toward the right side of the graphics window as shown.

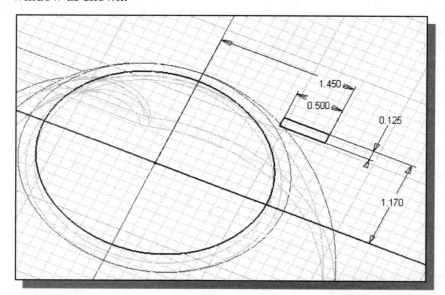

9. On your own, create and modify the dimensions as shown in the figure above.

10. Inside the graphics window, click once with the right-mouse-button to display the option menu. Select **Finish Sketch** in the popup menu to end the Sketch option.

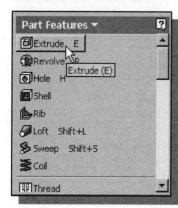

11. In the *Part Features* toolbar (the toolbar that is located to the left side of the graphics window), select the **Extrude** command by releasing the left-mouse-button on the icon.

12. Select the inside region of the rectangle as the profile of the extrusion.

13. Inside the *Extrude* dialog box, select the **Cut** operation and set the *Extents* **Distance** to **0.13 in** as shown.

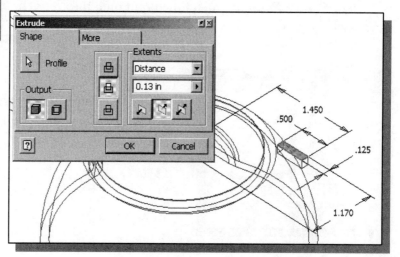

14. In the *Extrude* dialog box, click on the **OK** button to proceed with creating the cut feature.

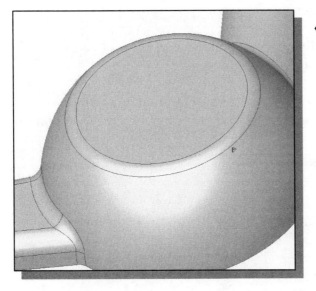

♦ Note that the pattern leader creates a fairly small cut on the solid model.

Creating a Rectangular Pattern

In *Autodesk Inventor*, existing features can be easily duplicated. The **Pattern** command allows us to create both rectangular and polar arrays of features. The patterned features are parametrically linked to the original feature; any modifications to the original feature are also reflected on the arrayed features.

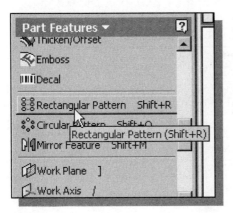

1. In the *Part Features* toolbar, select the **Rectangular Pattern** command by left-clicking once on the icon.

2. The message "*Select Feature to be arrayed:*" is displayed in the command prompt window. Select **Extrusion2**, the *cut feature created in the last section,* in the *Browser* window.

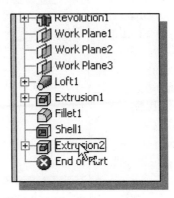

3. Inside the graphics window, right-mouse-click to bring up the option menu.

4. Select **Continue** in the option list to proceed with the Rectangular Pattern command.

5. In the *Rectangular Pattern* dialog box, notice the **selection** option for *Direction 1* is activated. Click on an edge along the swept feature as shown in the figure.

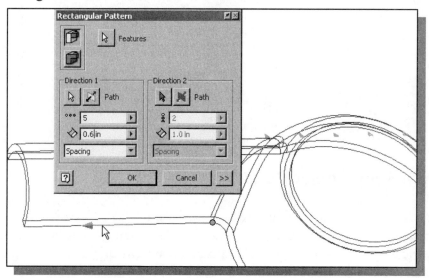

6. In the *Rectangular Pattern* dialog box, enter **5** in the *Count* box and **0.6** in the *Spacing* box as shown.

7. In the *Rectangular Pattern* dialog box, click on the **direction selection** icon for *Direction 2*.

8. Click on a straight edge along the lofted feature as shown in the figure.

9. In the *Rectangular Pattern* dialog box, enter **10** in the *Count* box and **0.25** in the *Spacing* box as shown.

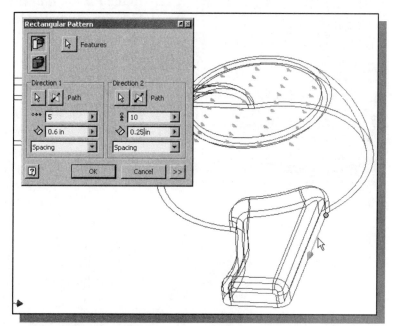

10. Click on the **OK** button to accept the settings and create the *rectangular pattern*.

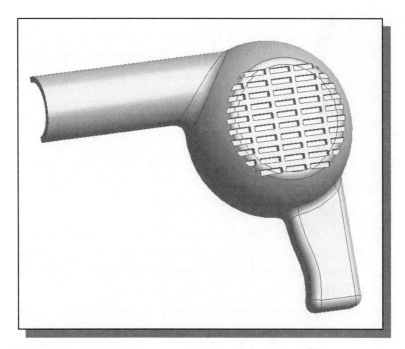

Creating a Swept Feature

❖ The **Sweep** operation is defined as moving a planar section through a planar(2D) or 3D path in space to form a three-dimensional solid object. The path can be an open curve or a closed loop, but must be on an intersecting plane with the profile. The **Extrusion** operation, which we have used in the previous lessons, is a specific type of sweep. The **Extrusion** operation is also known as a *linear sweep* operation, in which the sweep control path is always a line perpendicular to the two-dimensional section. Linear sweeps of unchanging shape result in what are generally called *prismatic solids* which means solids with a constant cross-section from end to end. In *Autodesk Inventor*, we create a *swept feature* by defining a path and then a 2D sketch of a cross section. The sketched profile is then swept along the planar path. The Sweep operation is used for objects that have uniform shapes along a trajectory.

◆ **Define a Sweep path**

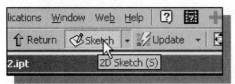

1. In the *Standard* toolbar select the **Sketch** command by left-clicking once on the icon.

2. In the *Status Bar* area, the message: "*Select face, workplane, sketch or sketch geometry*" is displayed. Select the *XZ Plane*, by left-clicking once on the *XZ Plane,* in the *Browser* window.

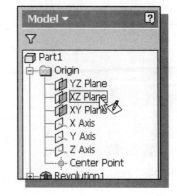

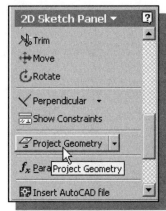

3. Select the **Project Geometry** command in the *2D Sketch Panel*.

4. In the *Status Bar* area, the message: "*Select face, workplane, sketch or sketch geometry*" is displayed. Select the outer edges of the bottom surface of the model as shown. (Hint: Use the Dynamic Zoom function to assist selecting all neighboring edges.)

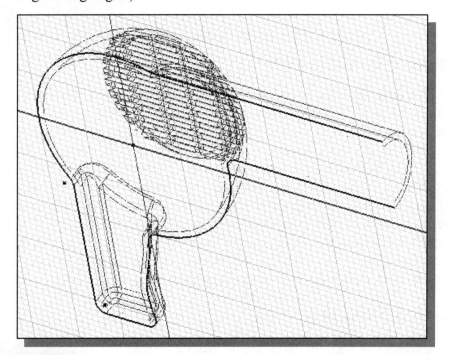

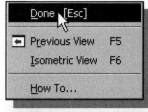

5. Inside the graphics window, right-mouse-click to bring up the option menu and select **Done** to end the Project Geometry command.

6. Inside the graphics window, click once with the right-mouse-button to display the option menu. Select **Finish Sketch** in the popup menu to end the Sketch option.

• The projected geometry will be used as the 2D sweep path of the feature.

♦ Define the Sweep Section

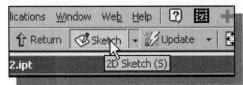

1. In the *Standard* toolbar select the **Sketch** command by left-clicking once on the icon.

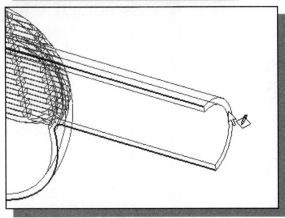

2. In the *Status Bar* area, the message: "*Select face, workplane, sketch or sketch geometry*" is displayed. Select the small circular surface of the model as shown.

3. Select the **Project Geometry** command in the *2D Sketch Panel*.

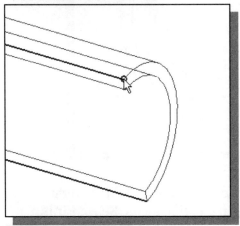

4. In the *Status Bar* area, the message: "*Select edge, vertex, work geometry or sketch geometry to project*" is displayed. Inside the graphics window, select the **top corner** of the circular surface to project onto the sketching plane.

5. Inside the graphics window, right-mouse-click to bring up the option menu and select **Done** to end the Project Geometry command.

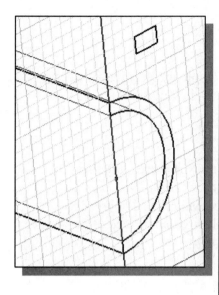

6. On your own, create a rectangle that is toward the top right side of the model as shown.

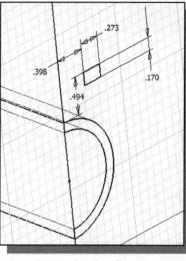

7. On your own, create the four dimensions to size and position the rectangle as shown. (Note that the values of the dimensions may be different than what is shown in the figure.)

8. Modify the dimensions as shown in the figure.

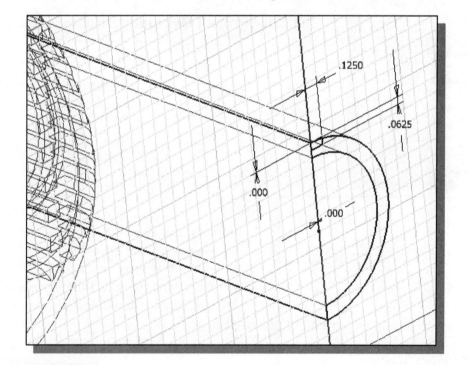

9. Inside the graphics window, click once with the right-mouse-button to display the option menu. Select **Finish Sketch** in the popup menu to end the Sketch option.

- The created 2D sketch will be used as the sweep section of the feature.

◆ **Completing the Swept Feature**

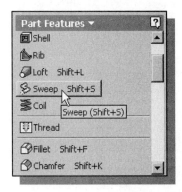

1. In the *Part Features* toolbar, select the **Sweep** command by left-clicking once on the icon.

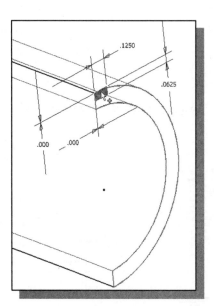

2. Select inside the rectangle region to use the rectangle as the sweep section.

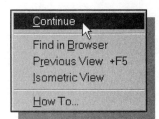

3. Right-mouse-click to bring up the option menu and select **Continue** in the option list to proceed with the command.

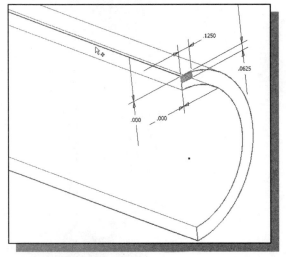

4. Click on the projected open curve we created as the sweep path as shown in the figure.

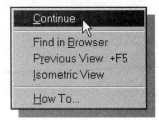

5. Inside the graphics window, right-mouse-click to bring up the option menu and select **Continue** in the option list to proceed with the command.

6. In the *Sweep* dialog box, confirm the operation to **Cut,** and then click on the
    OK button to accept the settings and create the *swept* feature.

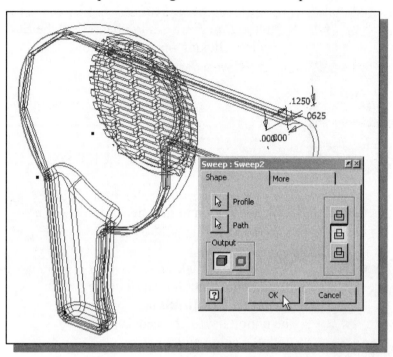

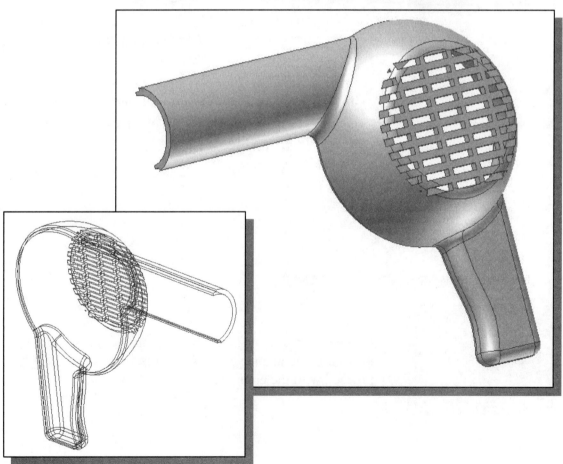

Questions:

1. Keeping the *History Tree* in mind, what is the difference between *cut with a pattern* and *cut each one individually*?

2. What is the difference between **Sweep** and **Extrude**?

3. What are the advantages and disadvantages of creating fillets using the **3D Fillets** command and creating fillets in the 2D profiles?

4. Describe the steps used to create the *Shell* feature in the lesson.

5. How do we modify the *Pattern* parameters after the model is built?

6. Describe the elements required in creating a *Swept* feature.

7. Create sketches showing the steps you plan to use to create the model shown on the next page:

Exercises:

1. Dimensions are in inches.

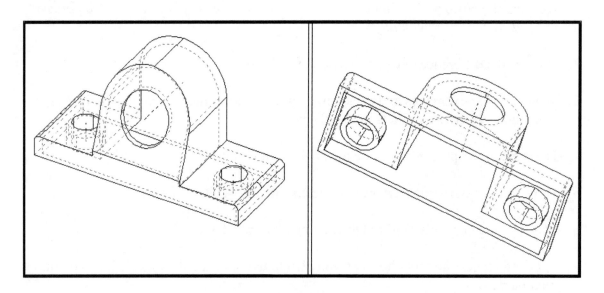

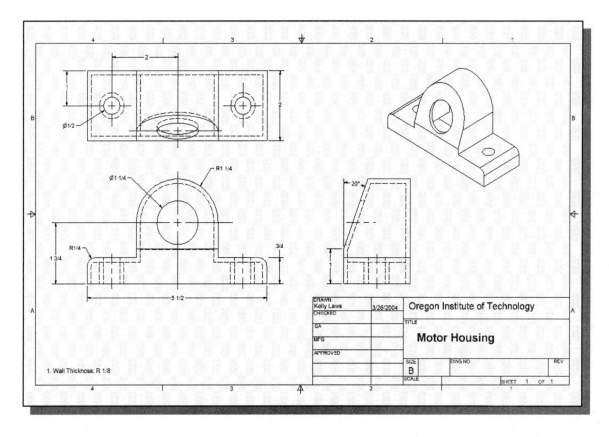

2. Dimensions are in inches.

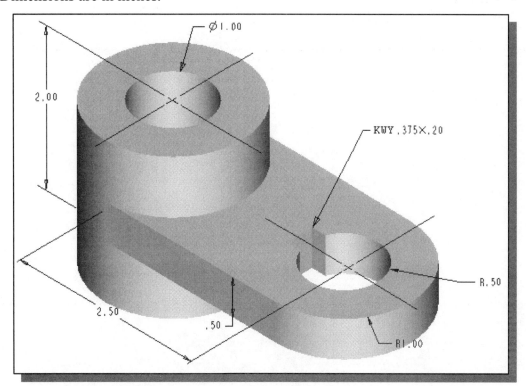

3. Dimensions are in inches.

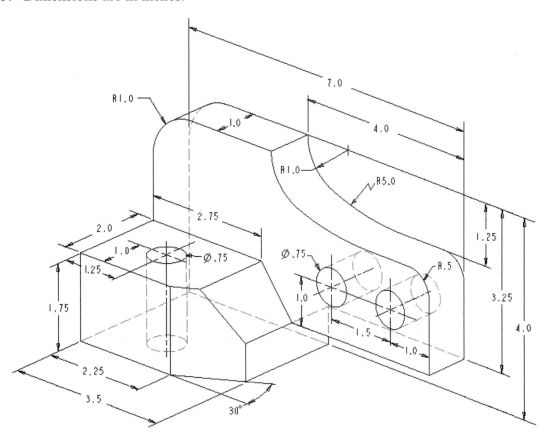

NOTES:

Chapter 12
Assembly Modeling – Putting It All Together

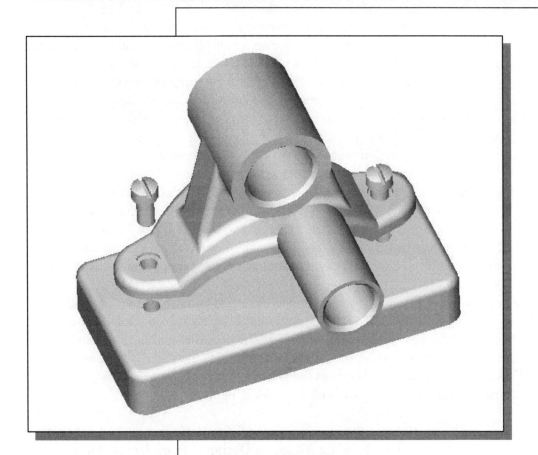

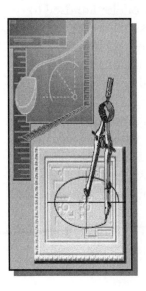

Learning Objectives

- ♦ **Understand the Assembly Modeling Methodology**
- ♦ **Create Parts in the Assembly Modeler Mode**
- ♦ **Understand and Utilize Assembly Constraints**
- ♦ **Understand the Autodesk Inventor DOF Display**
- ♦ **Utilize the Autodesk Inventor Adaptive Design Approach**
- ♦ **Create Exploded Assemblies**

Introduction

In the previous lessons, we have gone over the fundamentals of creating basic parts and drawings. In this lesson, we will examine the assembly modeling functionality of *Autodesk Inventor*. We will start with a demonstration on how to create and modify assembly models. The main task in creating an assembly is establishing the assembly relationships between parts. To assemble parts into an assembly, we will need to consider the assembly relationships between parts. It is a good practice to assemble parts based on the way they would be assembled in the actual manufacturing process. We should also consider breaking down the assembly into smaller subassemblies, which helps the management of parts. In *Autodesk Inventor*, a subassembly is treated the same way as a single part during assembling. Many parallels exist between assembly modeling and part modeling in parametric modeling software such as *Autodesk Inventor*.

Autodesk Inventor provides full associative functionality in all design modules, including assemblies. When we change a part model, *Autodesk Inventor* will automatically reflect the changes in all assemblies that use the part. We can also modify a part in an assembly. **Bi-directional full associative functionality** is the main feature of parametric solid modeling software that allows us to increase productivity by reducing design cycle time.

One of the key features of *Autodesk Inventor* is the use of an assembly-centric paradigm, which enables users to concentrate on the design without depending on the associated parameters or constraints. Users can specify how parts fit together and the *Autodesk Inventor assembly-based fit function* automatically determines the parts' sizes and positions. This unique approach is known as the **Direct Adaptive Assembly** approach, which defines part relationships directly with no order dependency.

In this lesson, we will also illustrate the basic concept of *Autodesk Inventor*'s **Adaptive Design** approach. The key element in doing **Adaptive Design** is to *underconstrain* features or parts. The applied *assembly constraints* in the assembly modeler are used to control the sizes, shapes, and positions of *underconstrained* sketches, features, and parts. No equations are required and this approach is extremely flexible when performing modifications and changes to the design. We can modify adaptive assemblies at any point, in any order, regardless of how the parts were originally placed or constrained.

In *Autodesk Inventor*, features and parts can be made adaptive at any time during creation or during assembly. The features of a part can be defined as adaptive when they are created in the part file. When we place such a part in an assembly, the features will then resize and change shape based on the applied assembly constraints. We can make features and parts adaptive from either the part modeling or assembly modeling environments.

The *Adaptive Design approach* is a unique design methodology that can only be found in *Autodesk Inventor*. The goal of this methodology is to improve the design process and allows you, the designer, to *design the way you think*.

Assembly Modeling Methodology

The *Autodesk Inventor* assembly modeler provides tools and functions that allow us to create 3D parametric assembly models. An assembly model is a 3D model with any combination of multiple part models. *Parametric assembly constraints* can be used to control relationships between parts in an assembly model.

Autodesk Inventor can work with any of the assembly modeling methodologies:

The Bottom Up approach
> The first step in the *bottom up* assembly modeling approach is to create the individual parts. The parts are then pulled together into an assembly. This approach is typically used for smaller projects with very few team members.

The Top Down approach
> The first step in the *top down* assembly modeling approach is to create the assembly model of the project. Initially, individual parts are represented by names or symbolically. The details of the individual parts are added as the project gets further along. This approach is typically used for larger projects or during the conceptual design stage. Members of the project team can then concentrate on the particular section of the project to which he/she is assigned.

The Middle Out approach
> The *middle out* assembly modeling approach is a mixture of the bottom-up and top-down methods. This type of assembly model is usually constructed with most of the parts already created and additional parts are designed and created using the assembly for construction information. Some requirements are known and some standard components are used, but new designs must also be produced to meet specific objectives. This combined strategy is a very flexible approach for creating assembly models.

The different assembly modeling approaches described above can be used as guidelines to manage design projects. Keep in mind that we can start modeling our assembly using one approach and then switch to a different approach without any problems.

In this lesson, the *bottom up* assembly modeling approach is illustrated. All of the parts (components) required to form the assembly are created first. *Autodesk Inventor's* assembly modeling tools allow us to create complex assemblies by using components that are created in part files or are placed in assembly files. A component can be a subassembly or a single part, where features and parts can be modified at any time. The sketches and profiles used to build part features can be fully or partially constrained. Partially constrained features may be adaptive, which means the size or shape of the associated parts are adjusted in an assembly when the parts are constrained to other parts. The basic concept and procedure of using the adaptive assembly approach is demonstrated in the tutorial.

The Shaft Support Assembly

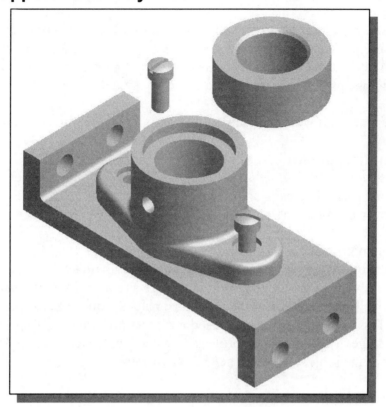

Additional Parts

- Four parts are required for the assembly: (1) **Collar**, (2) **Bearing**, (3) **Base-Plate** and (4) **Cap-Screw**. Create the four parts as shown below, then save the models as separate part files: *Collar*, *Bearing*, *Base-Plate*, and *Cap-Screw*. (Close all part files or exit *Autodesk Inventor* after you have created the parts.)

(1) *Collar*

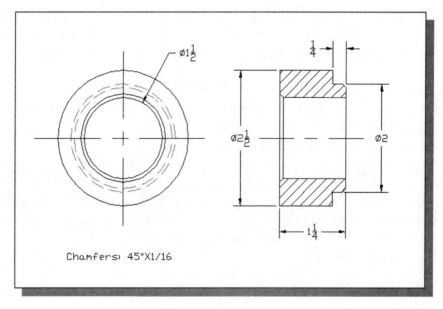

(2) **Bearing** (Construct the part with the datum origin aligned to the bottom center.)

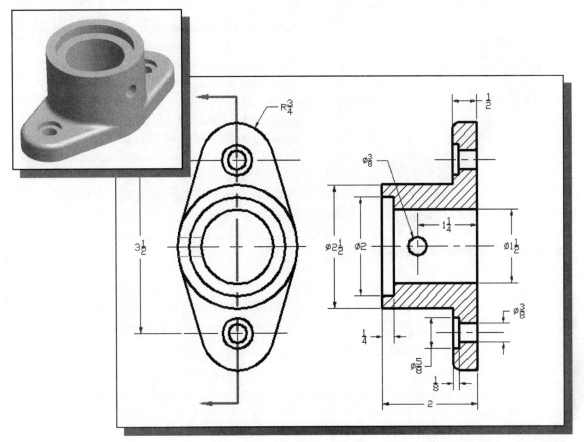

(3) **Base-Plate** (Construct the part with the datum origin aligned to the bottom center.)

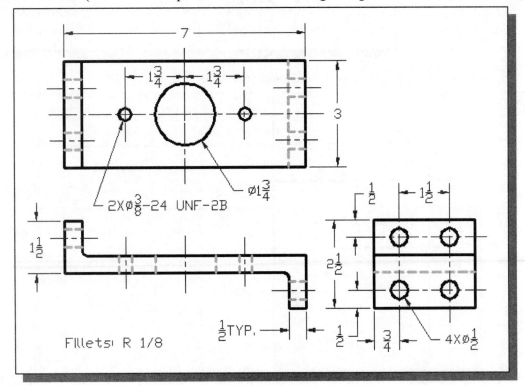

(4) *Cap-Screw*

- *Autodesk Inventor* provides two options for creating threads: **Thread** and **Coil**. The **Thread** command does not create true 3D threads; a pre-defined thread image is applied on the selected surface, as shown in the figure. The **Coil** command can be used to create true threads, which contain complex three-dimensional curves and surfaces. You are encouraged to experiment with the **Coil** command and/or the **Thread** command to create threads.

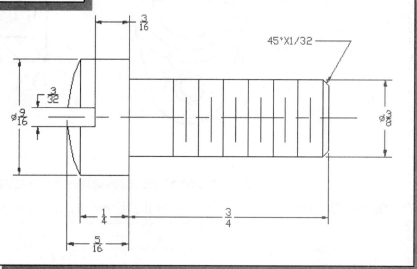

- Hint: First create a revolved feature using the profile shown below.

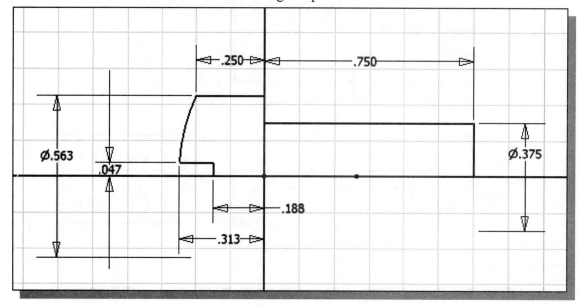

Starting *Autodesk Inventor*

1. Select the **Autodesk Inventor** option on the *Start* menu or select the **Autodesk Inventor** icon on the desktop to start *Autodesk Inventor*. The *Autodesk Inventor* main window will appear on the screen.

2. Once the program is loaded into memory, the ***Startup*** dialog box appears at the center of the screen.

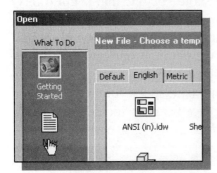

3. Select the **New** icon with a single click of the left-mouse-button in the *What to Do* dialog box.

4. Select the **English** tab and in the *New - Choose Template* area, select **Standard(in).iam** (Standard *Inventor* Assembly Model template file).

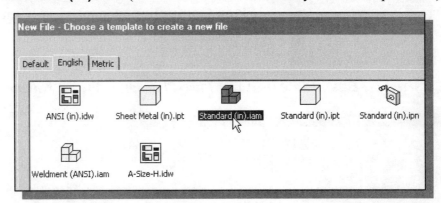

5. Click on the **OK** button in the *Startup* dialog box to accept the selected settings.

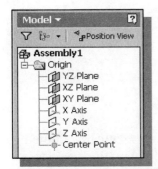

- In the *Browser* window, ***Assembly1*** is displayed with a set of work planes, work axes and a work point. In most aspects, the usage of work planes, work axes and work point is very similar to that of the *Inventor Part Modeler*.

- Notice, in the *Assembly Panel*, two component options are available: **Place Component** and **Create Component**. As the names imply, we can use parts that have been created or create new parts within the *Inventor Assembly Modeler*.

Placing the First Component

- The first component placed in an assembly should be a fundamental part or subassembly. The first component in an assembly file sets the orientation of all subsequent parts and subassemblies. The origin of the first component is aligned to the origin of the assembly coordinates and the part is grounded (all degrees of freedom are removed). The rest of the assembly is built on the first component, the **base component**. In most cases, this *base component* should be one that is **not likely to be removed** and **preferably a non-moving part** in the design. Note that there is no distinction in an assembly between components; the first component we place is usually considered as the *base component* because it is usually a fundamental component to which others are constrained. We can change the base component to a different base component by placing a new base component, specifying it as grounded, and then re-constraining any components placed earlier, including the first component. For our project, we will use the **Base-Plate** as the base component in the assembly.

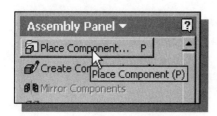

1. In the *Assembly Panel* (the toolbar that is located to the left side of the graphics window), select the **Place Component** command by left-mouse-clicking the icon.

2. Select the **Base-Plate** (part file: **Base-Plate.ipt**) in the list window.

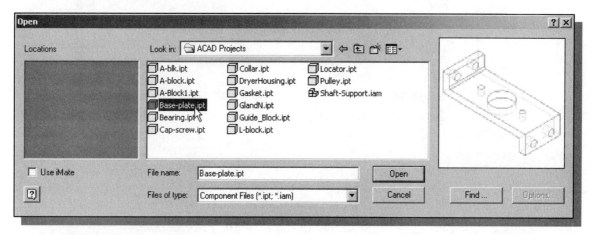

3. Click on the **Open** button to retrieve the model.

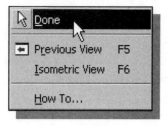

4. By default, the first component is automatically aligned to the origin of the assembly coordinates. We can also place multiple copies of the same component. Right-mouse-click once to bring up the option menu and select **Done** to end the placement of the *Base-Plate* part.

Placing the Second Component

➢ We will retrieve the *Bearing* part as the second component of the assembly model.

1. In the *Assembly Panel* (the toolbar that is located to the left side of the graphics window), select the **Place Component** command by left-mouse-clicking the icon.

2. Select the ***Bearing*** design (part file: ***Bearing.ipt***) in the list window. And click on the **Open** button to retrieve the model.

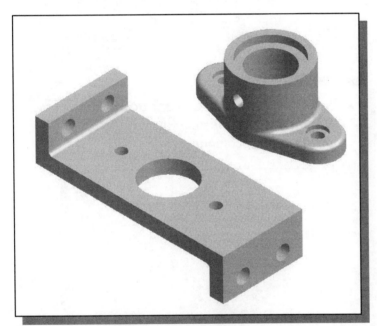

3. Place the *Bearing* toward the upper right corner of the graphics window, as shown in the figure.

4. Inside the graphics window, right-mouse-click once to bring up the option menu and select **Done** to end the placement of the *Bearing* part.

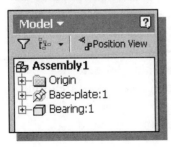

• Inside the *Browser* window, the retrieved parts are listed in their corresponding order. The **Pin** icon in front of the *Base-Plate* filename signifies the part is grounded and all *six degrees of freedom* are restricted. The number behind the filename is used to identify the number of copies of the same component in the assembly model.

Degrees of Freedom and Constraints

- Each component in an assembly has six **degrees of freedom** (**DOF**), or ways in which rigid 3D bodies can move: movement along the X, Y, and Z axes (translational freedom), plus rotation around the X, Y, and Z axes (rotational freedom). *Translational DOF*s allow the part to move in the direction of the specified vector. *Rotational DOF*s allow the part to turn about the specified axis.

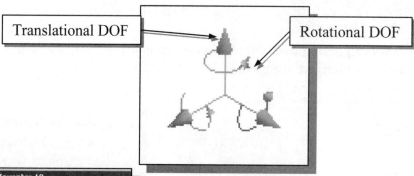

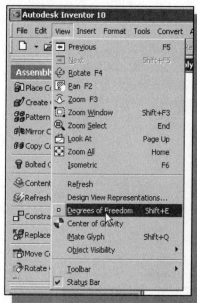

> Select the **Degrees Of Freedom** option in the **View** pull-down menu to display the DOF of the unconstrained component.

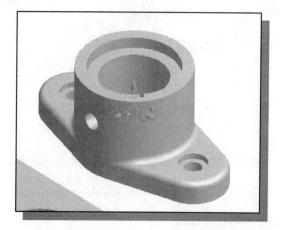

- In *Autodesk Inventor*, the degrees-of-freedom symbol shows the remaining degrees of freedom (both translational and rotational) for one or more components of the active assembly. When a component is fully constrained in an assembly, the component cannot move in any direction. The position of the component is fixed relative to other assembly components. All of its degrees of freedom are removed. When we place an assembly constraint between two selected components, they are positioned relative to one another. Movement is still possible in the unconstrained directions.

> It is usually a good idea to fully constrain components so that their behavior is predictable as changes are made to the assembly. Leaving some degrees of freedom open can sometimes help retain design flexibility. As a general rule, we should use only enough constraints to ensure predictable assembly behavior and avoid unnecessary complexity.

Assembly Constraints

• We are now ready to assemble the components together. We will start by placing assembly constraints on the **Bearing** and the **Base-Plate**.

To assemble components into an assembly, we need to establish the assembly relationships between components. It is a good practice to assemble components the way they would be assembled in the actual manufacturing process. **Assembly constraints** create a parent/child relationship that allows us to capture the design intent of the assembly. Because the component that we are placing actually becomes a child to the already assembled components, we must use caution when choosing constraint types and references to make sure they reflect the intent.

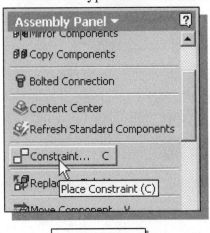

➢ In the *Assembly Panel,* select the **Constraint** command by left-mouse-clicking once on the icon.

• The *Place Constraints* dialog box appears on the screen. Four types of assembly constraints are available.

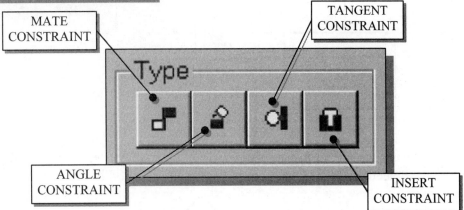

• Assembly models are created by applying proper *assembly constraints* to the individual components. The constraints are used to restrict the movement between parts. Constraints eliminate rigid body degrees of freedom (**DOF**). A 3D part has *six degrees of freedom* since the part can rotate and translate relative to the three coordinate axes. Each time we add a constraint between two parts, one or more DOF is eliminated. The movement of a fully constrained part is restricted in all directions. Four basic types of assembly constraints are available in *Autodesk Inventor*: **Mate, Angle, Tangent** and **Insert**. Each type of constraint removes different combinations of rigid body degrees of freedom. Note that it is possible to apply different constraints and achieve the same results.

➢ **Mate** – Constraint positions components face-to-face, or adjacent to one another, with faces flush. Removes one degree of linear translation and two degrees of angular rotation between planar surfaces. Selected surfaces point in opposite directions and can be **offset** by a specified distance. Mate constraint positions selected faces normal to one another, with faces coincident.

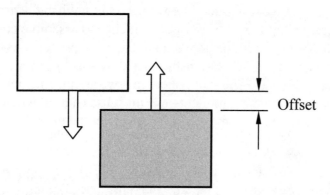

➢ **Flush** – Makes two planes coplanar with their faces aligned in the same direction. Selected surfaces point in the same direction and are offset by a specified distance. Flush constraint aligns components adjacent to one another with faces flush and positions selected faces, curves, or points so that they are aligned with surface normals pointing in the same direction. (Note that the Flush constraint is listed as a selectable option in the Mate constraint.)

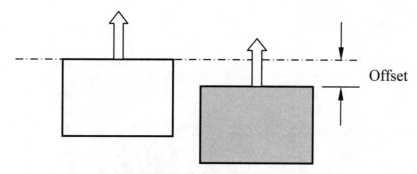

➢ **Angle** – Creates an angular assembly constraint between parts, subassemblies, or assemblies. Selected surfaces point in the direction specified by the angle.

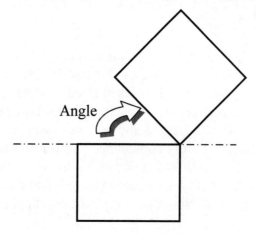

➢ **Tangent** – Aligns selected faces, planes, cylinders, spheres, and cones to contact at the point of tangency. Tangency may be on the inside or outside of a curve, depending on the selection of the direction of the surface normal. A Tangent constraint removes one degree of translational freedom.

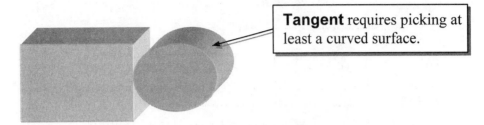

Tangent requires picking at least a curved surface.

➢ **Insert** – Aligns two circles, including their center axes and planes. Selected circular surfaces become co-axial. Insert constraint is a combination of a face-to-face Mate constraint between planar faces and a Mate constraint between the axes of the two components. A rotational degree of freedom remains open. The surfaces do not need to be full 360-degree circles. Selected surfaces can point in opposite directions or in same direction and can be offset by a specified distance.

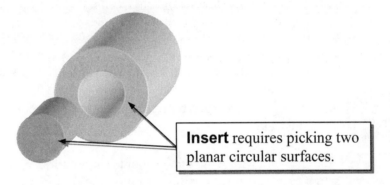

Insert requires picking two planar circular surfaces.

Apply the First Assembly Constraint

1. In the *Place Constraint* dialog box, confirm the constraint type is set to **Mate** constraint and select the top horizontal surface of the base part as the first part for the **Mate** alignment command.

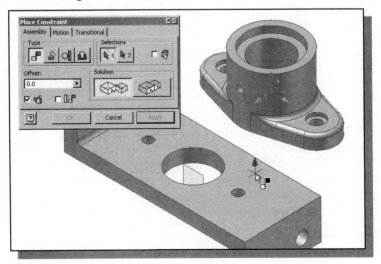

2. On your own, dynamically rotate the displayed model to view the bottom of the *Bearing* part, as shown in the figure below.

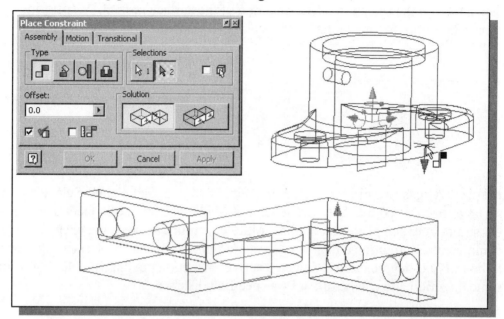

3. Click on the bottom face of the *Bearing* part as the second part selection to apply the constraint. Note the direction normals shown in the figure; the Mate constraint requires the selection of opposite direction of surface normals.

4. Click on the **Apply** button to accept the selection and apply the Mate constraint.

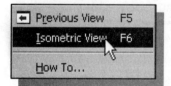

5. Move the cursor inside the graphics window, and then right-mouse-click to bring up the option menu.

6. Select the **Isometric View** to adjust the display of the assembly model.

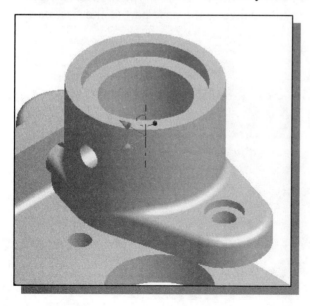

❖ Notice the **DOF** symbol is adjusted automatically in the graphics window. The Mate constraint removes one degree of linear translation and two degrees of angular rotation between the selected planar surfaces. The *Bearing* part can still move along two axes and rotate about the third axis.

Apply a Second MATE Constraint

❖ The **Mate** constraint can also be used to align axes of cylindrical features.

1. In the *Place Constraint* dialog box, confirm the constraint type is set to **Mate** constraint and the *Offset* option is set to **0.00**.

2. Move the cursor near the cylindrical surface of the right counter bore hole of the *Bearing* part. Select the axis when it is displayed as shown. (Hint: Use the *Dynamic Rotation* option to assist the selection.)

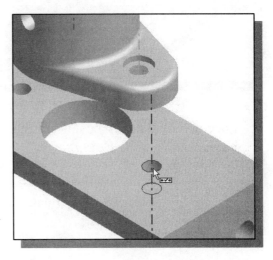

3. Move the cursor near the cylindrical surface of the small hole on the *Base-Plate* part. Select the axis when it is displayed as shown.

4. In the *Place Constraint* dialog box, click on the **Apply** button to accept the selection and apply the **Mate** constraint.

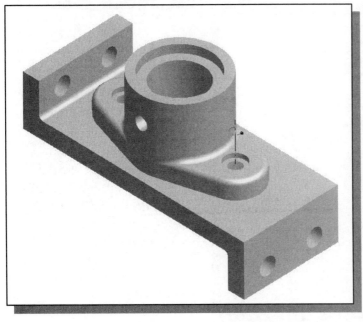

5. In the *Place Constraint* dialog box, click on the **Cancel** button to exit the Place Constraint command.

❖ The *Bearing* part appears to be placed in the correct position. But the **DOF** symbol indicates that this is not the case; the bearing part can still rotate about the displayed vertical axis.

Constrained Move

❖ To see how well a component is constrained, we can perform a constrained move. A constrained move is done by dragging the component in the graphics window with the left-mouse-button. A constrained move will honor previously applied assembly constraints. That is, the selected component and parts constrained to the component move together in their constrained positions. A grounded component remains grounded during the move.

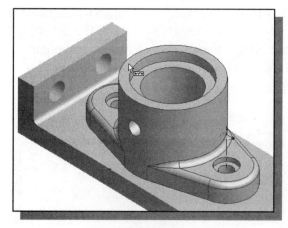

1. Inside the graphics window, move the cursor on top of the top surface of the *Bearing* part as shown in the figure.

2. Press and hold down the left-mouse-button and drag the *Bearing* part downward.

❖ The *Bearing* part can freely rotate about the displayed axis.

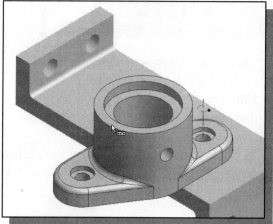

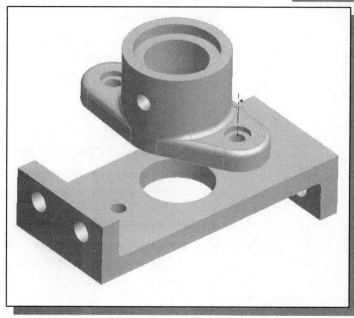

3. On your own, use the dynamic rotation command to view the alignment of the *Bearing* part.

4. Rotate the *Bearing* part and adjust the display as shown in the figure.

5. Switch to the wireframe display mode before proceeding to the next section.

Apply a Flush Constraint

- Besides selecting the surfaces of solid models to apply constraints, we can also select the established work planes to apply the assembly constraints. This is an additional advantage of using the *BORN technique* in creating part models. For the *Bearing* part, we will apply a Mate constraint to two of the work planes and eliminate the last rotational DOF.

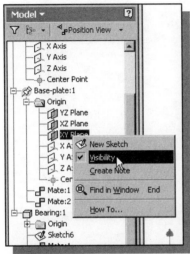

1. On your own, inside the *Browser* window toggle *ON* the **Visibility** for the corresponding work planes that can be used for alignment of the *Base-Plate* and the *Bearing* parts.

2. In the *Assembly Panel,* select the **Constraint** command by left-mouse-clicking once on the icon.

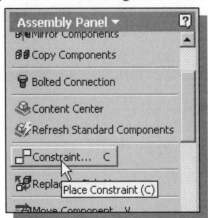

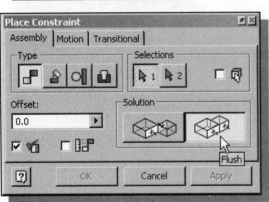

3. In the *Place Constraint* dialog box, confirm the constraint type is set to **Mate** constraint and the *Offset* option is set to **0.00**.

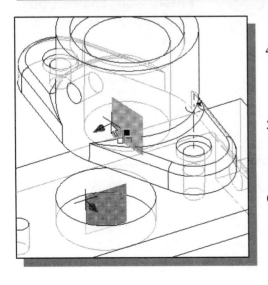

4. In the *Place Constraint* dialog box, switch the *Solution* option to **Flush** as shown.

5. Select the *work plane* of the *Base-Plate* part as the first part for the Flush alignment command.

6. Select the corresponding *work plane* of the *Bearing* part as the second part for the Flush alignment command.

❖ Note that the Flush constraint makes two planes coplanar with their faces aligned in the same direction.

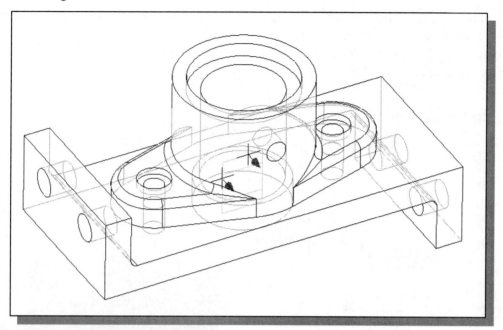

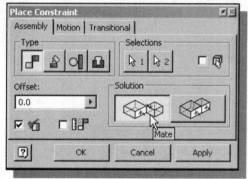

7. In the *Place Constraint* dialog box, switch the *Solution* option to **Mate** and notice the *Bearing* part is rotated 180 degrees to satisfy the **Mate** constraint.

❖ Note that the *Show Preview* option allows us to preview the result before accepting the selection.

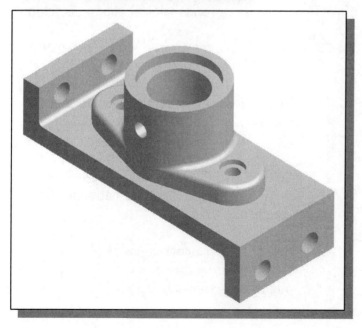

8. On your own, switch to the **Flush** option and click on the **Apply** button to accept the settings.

9. In the *Place Constraint* dialog box, click on the **Cancel** button to exit the Place Constraint command.

❖ Note the **DOF** symbol disappears, which indicates the assembly is fully constrained.

Placing the Third Component

➤ We will retrieve the *Collar* part as the third component of the assembly model.

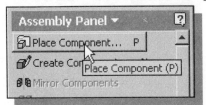

1. In the *Assembly Panel* (the toolbar that is located to the left side of the graphics window), select the **Place Component** command by left-mouse-clicking the icon.

2. Select the ***Collar*** design (part file: ***Collar.ipt***) in the list window. And click on the **Open** button to retrieve the model.

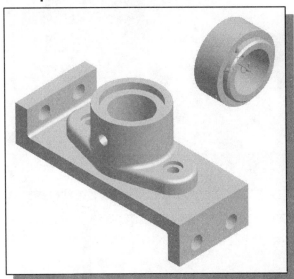

3. Place the *Collar* part toward the upper right corner of the graphics window, as shown in the figure.

4. Inside the graphics window, right-mouse-click once to bring up the option menu and select **Done** to end the placement of the *Collar* part.

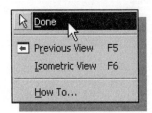

❖ Notice the DOF symbol displayed on the screen. The *Collar* part can move linearly and rotate about the three axes (six degrees of freedom).

Applying an Insert Constraint

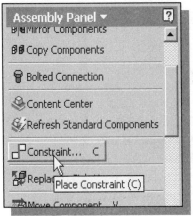

1. In the *Assembly Panel,* select the **Place Constraint** command by left-mouse-clicking once on the icon.

2. In the *Place Constraint* dialog box, switch to the **Insert** constraint.

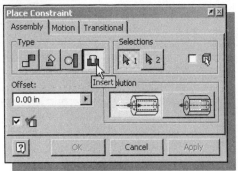

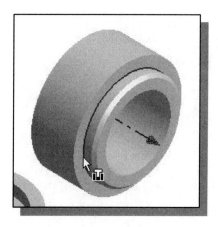

3. Select the inside corner of the *Collar* part as the first surface to apply the Insert constraint, as shown in the figure.

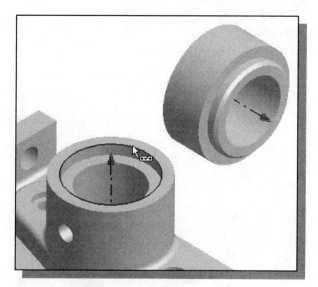

4. Select the inside circle on the top surface of the *Bearing* part as the second surface to apply the Insert constraint, as shown in the figure.

5. Click on the **Apply** button to accept the settings.

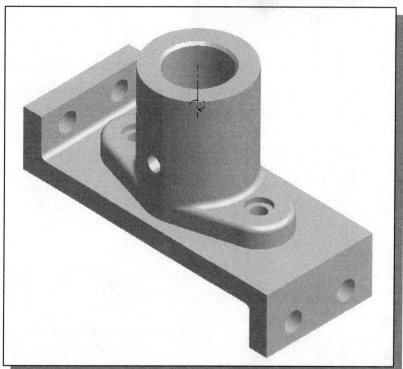

➢ Note that one rotational degree of freedom remains open; the *Collar* part can still freely rotate about the displayed DOF axis.

Assemble the Cap-Screws

❖ We will place two *Cap-Screw* parts to complete the assembly model.

1. In the *Assembly Panel* (the toolbar that is located to the left side of the graphics window), select the **Place Component** command by left-mouse-clicking once on the icon.

2. Select the **Cap-Screw** design (part file: ***Cap-Screw.ipt***) in the list window. And click on the **Open** button to retrieve the model.

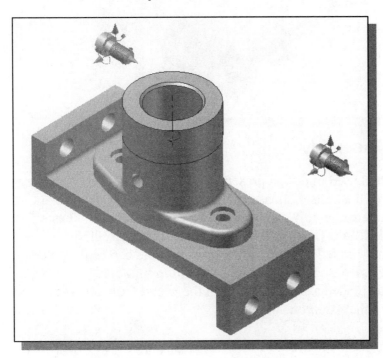

3. Place two copies of the *Cap-Screw* part on both sides of the *Collar* by clicking twice on the screen as shown in the figure.

4. Inside the graphics window, right-mouse-click once to bring up the option menu and select **Done** to end the Place Component command.

❖ Notice the DOF symbols displayed on the screen. Each *Cap-Screw* has six degrees of freedom. Both parts are referencing the same external part file, but each can be constrained independently.

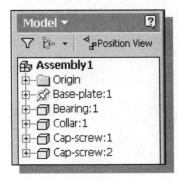

• Inside the *Browser* window, the retrieved parts are listed in the order they are placed. The number behind the part name is used to identify the number of copies of the same part in the assembly model. Move the cursor to the last part name and notice the corresponding part is highlighted in the graphics window.

> ➢ On your own, use the **Place Constraints** command and assemble the *Cap-Screws* in place as shown in the figure below.

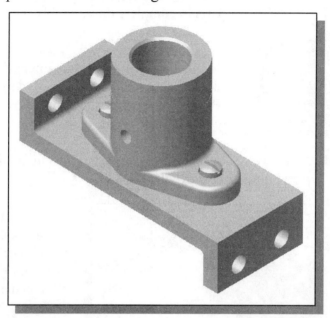

Exploded View of the Assembly

- Exploded assemblies are often used in design presentations, catalogs, sales literature, and in the shop to show all of the parts of an assembly and how they fit together. In *Autodesk Inventor*, an exploded assembly can be created by two methods: (1) using the **Move Component** and **Rotate Component** commands in the *Assembly Modeler*, which contains only limited options for the operation but can be done very quickly; (2) transferring the assembly model into the *Presentation Modeler*. For our example, we will create an exploded assembly by using the **Move Component** command that is available in the *Assembly Modeler*.

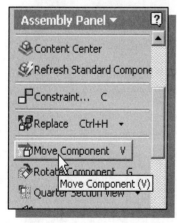

1. In the *Assembly Panel,* select the **Move Component** command by left-mouse-clicking once on the icon.

2. Inside the graphics window, move the cursor on top of the top surface of the *Collar* part as shown in the figure.

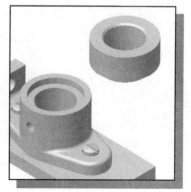

3. Press and hold down the left-mouse-button and drag the *Collar* part toward the right side of the assembly as shown.

4. On your own, repeat the above steps and create an exploded assembly by repositioning the components as shown in the figure below.

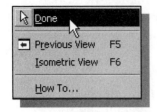

5. Inside the graphics window, right-mouse-click once to bring up the option menu and select **Done** to end the Move Component command.

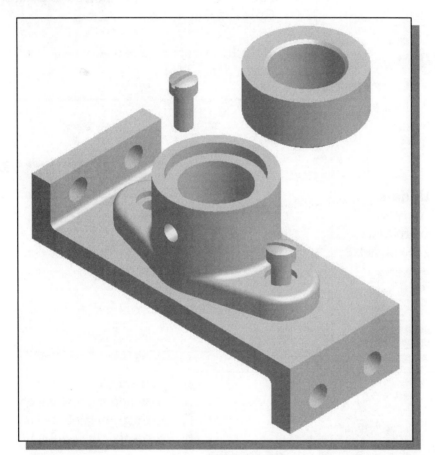

❖ The Move Component and Rotate Component commands are used to temporarily reposition the components in the graphics window. The displayed image is temporary, but it can be printed with the **Print** command through the pull-down menu.

6. Click on the **Update** button in the *Standard* toolbar area.

❖ Note that the components are reset back to their assembled positions, based on the applied assembly constraints.

Editing the Components

- The *associative functionality* of *Autodesk Inventor* allows us to change the design at any level, and the system reflects the changes at all levels automatically.

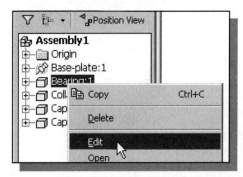

1. Inside the *Desktop Browser*, move the cursor on top of the **Bearing** part. Right-mouse-click once to bring up the option menu and select **Edit** in the option list.

❖ Note that we are automatically switched back to *Part Editing Mode*.

2. On your own, adjust the diameter of the small *Drill Hole* to **0.25** as shown.

3. Click on the **Update** button in the *Standard* toolbar area to proceed with updating the model.

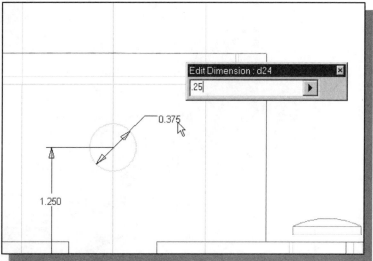

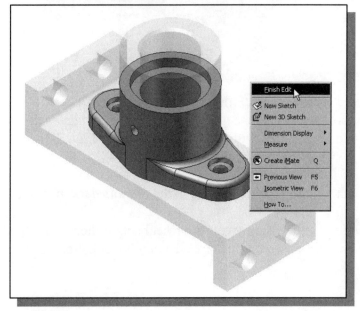

4. Inside the graphics window, click once with the right-mouse-button to display the option menu.

5. Select **Finish Edit** in the popup menu to exit *Part Editing Mode* and return to *Assembly Mode*.

➢ *Autodesk Inventor* has updated the part in all levels and in the current *Assembly Mode*. On your own, open the **Bearing** part to confirm the modification is performed.

Adaptive Design Approach

- *Autodesk Inventor*'s **Adaptive Design** approach allows us to use the applied *assembly constraints* to control the sizes, shapes, and positions of **underconstrained** sketches, features, and parts.

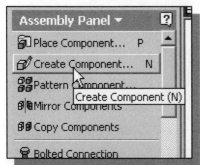

1. In the *Assembly Panel,* select the **Create Component** command by left-mouse-clicking once on the icon.

2. In the *Create In-Place Component* dialog box, enter **A-Block** as the new file name.

3. Click on the **OK** button to accept the settings.

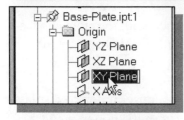

4. Click on the *XY Plane* of the *Base-Plate* part, inside the *Browser* window, to apply a **Flush** constraint.

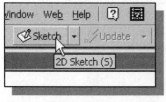

5. In the *Standard* toolbar select the **Sketch** command by left-clicking once on the icon.

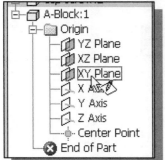

6. Select the *XY Plane*, in the browser window, of the *A-Block* part to align the sketch plane.

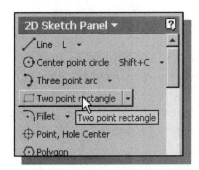

7. Select the **Two point rectangle** command by clicking once with the left-mouse-button on the icon in the *2D Sketch Panel*.

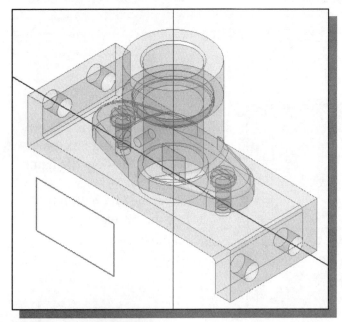

8. On your own, create a rectangle of arbitrary size below the assembly model as shown in the figure.

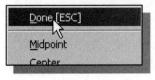

9. Inside the graphics window, click once with the right-mouse-button to display the option menu and select **Done** to end the Rectangle command.

10. Inside the graphics window, click once with the right-mouse-button to display the option menu. Select **Create Feature → Extrude** in the popup menu.

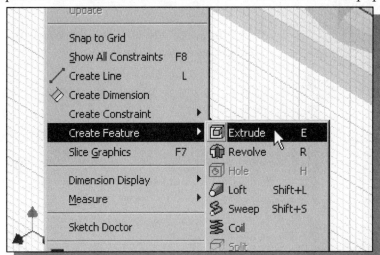

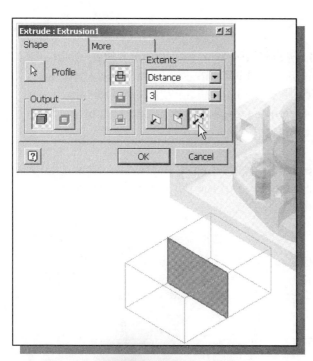

11. Inside the *Extrude* dialog box, select the **Both Sides** option and set the *Extents Distance* to **3 in** as shown.

12. In the *Extrude* dialog box, click on the **OK** button to proceed with creating the feature.

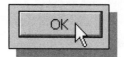

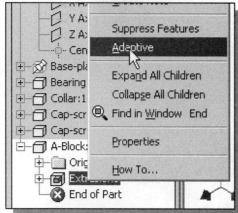

13. Right-mouse-click on the ***Extrusion1*** feature of the *A-Block* part, inside the *Browser window,* and select **Adaptive** to allow the use of the adaptive design approach.

14. Click inside the graphics window to deselect any part or assembly.

15. Inside the graphics window, click once with the right-mouse-button to display the option menu.

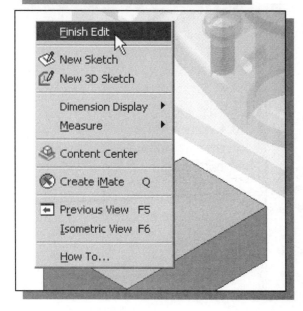

16. Select **Finish Edit** in the popup menu to exit *Part Editing Mode* and return to *Assembly Modeling Mode.*

- Note that the rectangular block is created with only one dimension, the extrude distance. The 2D sketch of the part is intentionally underconstrained.

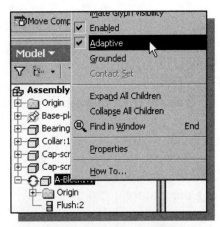

17. Right-mouse-click on the *A-Block* part, inside the *Browser* window, and confirm the **Adaptive** option is turned *ON*. This option allows the use of the *Adaptive Design* approach in the current assembly model.

- Note the **Adaptive** icon, the two-arrow symbol, appears in front of the part name in the *Browser* window.

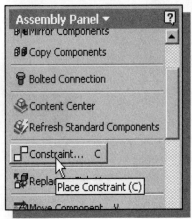

18. In the *Assembly Panel*, select the **Constraint** command by left-mouse-clicking once on the icon.

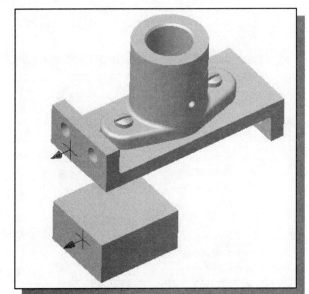

19. On your own, use the **Flush** constraint to align the *A-block* to the left-edge of the *Base-Plate* as shown below.

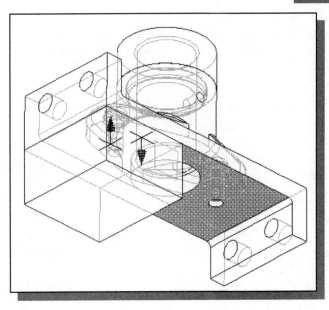

20. Create a **Mate** constraint to align the top of the *A-block* to the bottom of the *Base-Plate* part as shown.

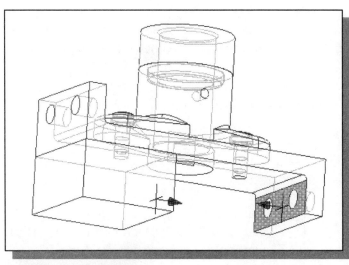

21. Use the **Mate** constraint and align the right surface of the *A-block* to the bottom left surface of the *Base-Plate* part as shown.

- Note that the length of the *A-block* part is automatically adjusted to fit the defined constraint.

22. In the *Place Constraint* dialog box, click on the **Cancel** button to end the **Place Constraint** command.

Delete and Re-apply Assembly Constraints

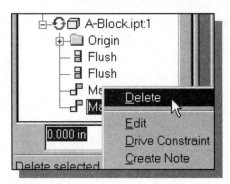

1. Inside the *Browser* window, right-mouse-click on the last **Mate** constraint of the *A-Block* part to bring up the option menu.

2. Select **Delete** to remove the applied constraint.

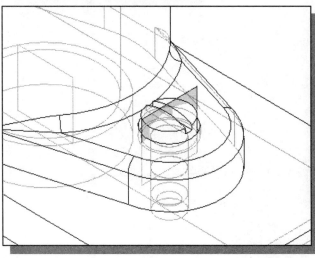

3. On your own, switch *ON* the **Visibility** of the vertical work plane of the *Cap-Screw* part, which is perpendicular to the length of the *Base-Plate* part, as shown.

4. In the *Assembly Panel*, select the **Place Constraint** command by left-mouse-clicking once on the icon.

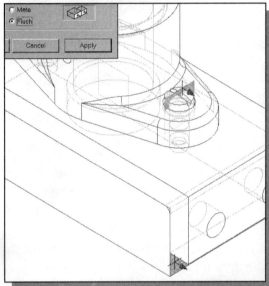

5. On your own, align the right vertical surface of the *A-block* part to the vertical work plane as shown in the figure.

6. On your own, experiment with aligning the right vertical surface of the *A-block* part to other vertical surfaces of the assembly model.

• As can be seen, the length of the *A-block* is adjusted to the newly applied constraint. The ***Adaptive Design*** approach allows us to have greater flexibility and simplifies the design process.

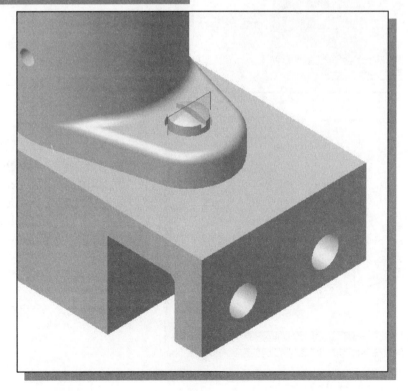

Setup a Drawing of the Assembly Model

1. Click on the **drop-down arrow** next to the **New File** icon in the *Standard* toolbar area to display the available New File options.

2. Select **Drawing** from the option list.

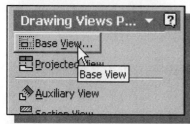

3. Click on the **Base View** in the *Drawing Views Panel* to create a base view.

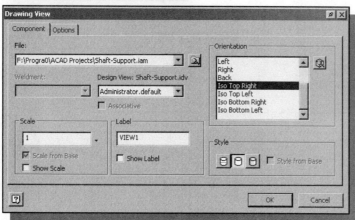

4. In the *Drawing View* dialog box, set *Orientation* to **Iso Top Right View** and **Hidden Line** as shown in the figure. (**DO NOT** click on the **OK** button at this point.)

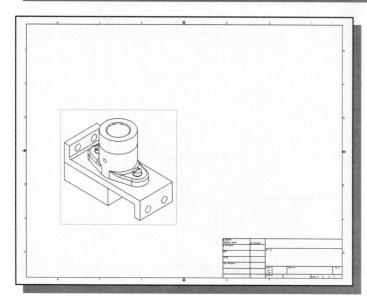

5. Move the cursor inside the graphics window and place the ***Base*** view near the lower left side of the *Border* as shown.

❖ Note that the default sheet size is much bigger than the created view. *Inventor* allows us to adjust the sheet size even when views have been created.

6. Inside the *Drawing Browser* window, right-mouse-click on **Sheet:1** to display the option menu.

7. Select **Edit Sheet** in the option menu to display the settings for the drawing.

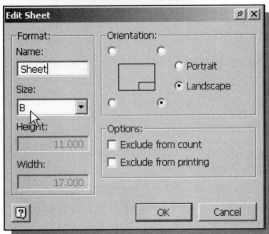

8. Set the sheet size to **B-size** as shown.

9. Click on the **OK** button to accept the settings and exit the **Edit Sheet** command.

❖ Note that the *Border* and *Title Block* are automatically replaced as the sheet size is adjusted.

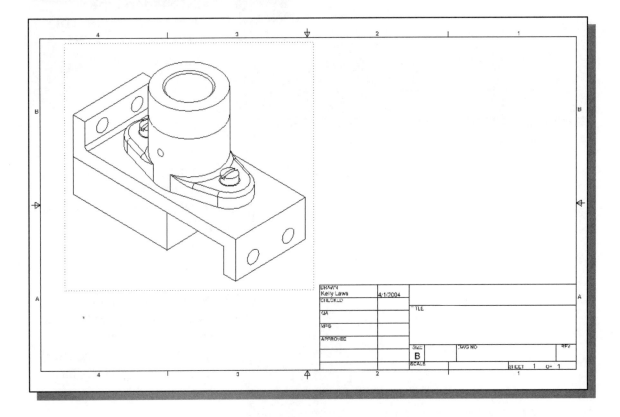

Creating a Parts List

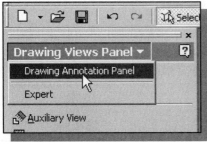

1. Left-mouse-click in the *title* area of the *Drawing Views Panel*.

2. Select **Drawing Annotation Panel** by left-clicking once in the drop-down list.

3. In the *Drawing Annotation* window, click on the **Parts List** button.

4. In the prompt area, the message "*Select a view*" is displayed. Items in the selected view will be listed in the *Parts List*. Select the base view.

❖ The *Parts List* dialog box appears; options are available to make adjustments to the numbering system and table wrap settings

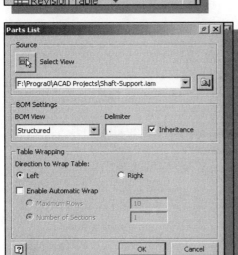

➢ The *Parts List* – BOM options are as follows:

Structured: Creates a parts list in which subassemblies are assigned using a nested numbering system (for example, 1, 1.1, 1.1.1). The nested number extends as many levels as needed for the assembly levels in the model.

Only Parts: Creates a parts list that sequentially numbers all parts in the assembly, including parts that are contained in subassemblies.

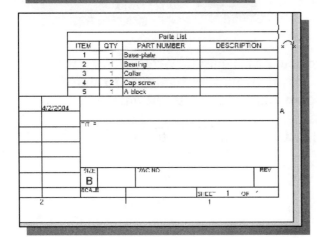

❖ Note that no subassemblies are used in the assembly; we will accept the default settings.

5. Click **OK** to accept the default settings.

6. Place the *Parts List* above the *Title Block* as shown.

Editing the Parts List

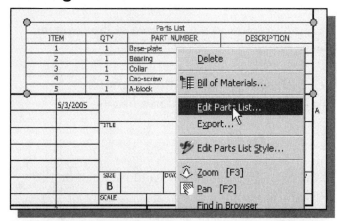

1. Move the cursor on top of the *Parts List* and right-mouse-click once to display the option menu.

2. Choose **Edit Parts List** in the option menu as shown.

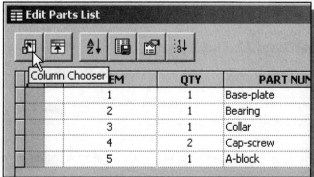

3. Click on the **Column Chooser** button as shown.

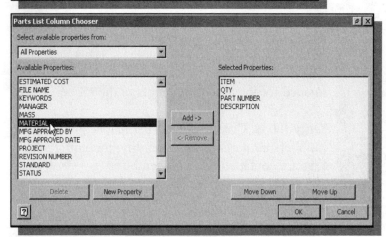

4. Select **MATERIAL** in the *Available Properties* list as shown.

5. Click **Add** to add the selected item to the *Selected Properties* list.

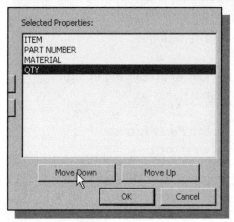

6. On your own, adjust the *Selected Properties* list as shown. (Hint: Use the **Move Up** and **Move Down** buttons to arrange the order of the list.)

7. Click **OK** to accept the settings.

❖ The *Parts List* is adjusted using the new settings. Note that, currently, all of the parts are using the **Default** material type.

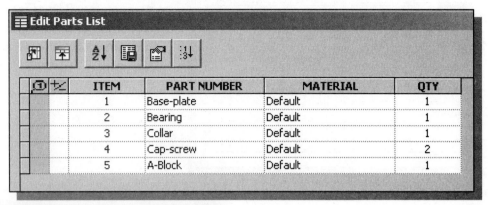

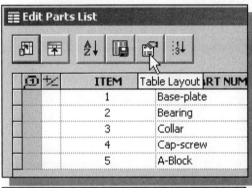

8. Click on the **Table Layout** button as shown.

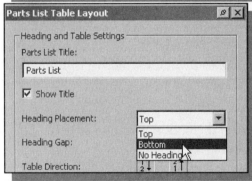

9. Set the *Heading Placement* to **Bottom** as shown.

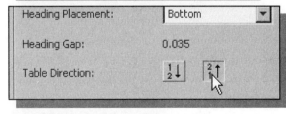

10. Set the *Table Direction* to **Add new parts to top** as shown.

11. On your own, clear the *Parts List Title*.

12. Click **OK** to accept the settings.

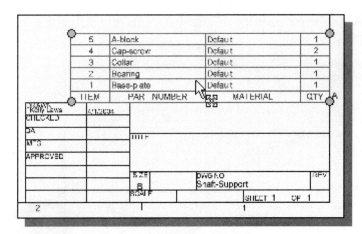

13. On your own, adjust the position the **Parts List** so that it is aligned to the top edge of the *Title Block* as shown.

Changing the Material Type

❖ We will switch back to the assembly model to change the assignments of the material type.

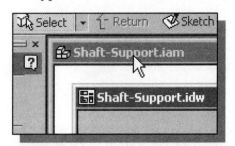

1. Click on the *Shaft-Support.iam* window to switch back to the assembly model.

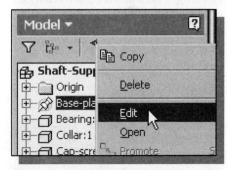

2. Inside the *Model Browser* window, right-mouse-click on **Base-Plate:1** to display the option menu.

3. Select **Edit** in the option menu to enter the *Edit Mode* for the selected part.

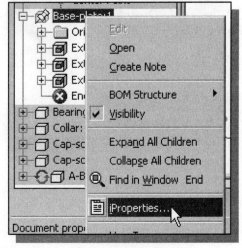

4. Inside the *Model Browser* window, select the **Base-plate** part by clicking once with the left-mouse-button.

5. Right-mouse-click on **Base-plate:1** to display the option menu and choose **iProperties** in the options list.

6. Click on the **Physical** tab in the *Base-plate.ipt Properties* window.

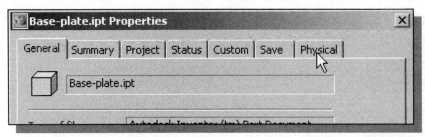

7. Choose the **Steel, Mild** in the *Material* list. Note the properties of the selected material are also displayed in the *Properties* list as shown in the figure.

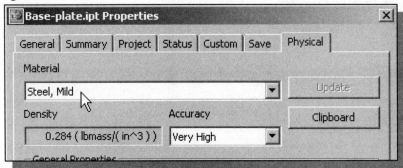

❖ *Autodesk Inventor* comes with many pre-entered material information, additional material types/properties can also be added/changed as well.

8. Click **OK** to accept the setting and exit the *Materials* dialog box.

9. Inside the graphics window, click once with the right-mouse-button to display the option menu.

10. Select **Finish Edit** in the popup menu to exit the *Part Editing Mode*.

11. On your own, switch back to the *Shaft-Support.idw* window and notice the *Material* information for the **Base-plate** part is now updated

12. On your own, repeat the above steps to change the material information for the other parts as shown in the figure below.

ITEM	PART NUMBER	MATERIAL	QTY
5	A-block	Aluminum-6061	1
4	Cap-screw	Steel, Mild	2
3	Collar	Bronze, Soft Tin	1
2	Bearing	Cast Iron	1
1	Base-plate	Steel, Mild	1

DRAWN Randy	5/3/2005		A
CHECKED			

Completing the Assembly Drawing

1. In the *Drawing Annotation* window, click on the **Balloon** button.

2. In the prompt area, the message "*Select a location*" is displayed. Click on the *Collar* part to attach an arrowhead to the part.

3. Pick another location to place the balloon as shown in the figure.

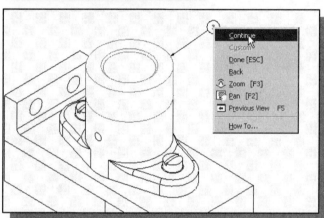

4. Inside the graphics window, click once with the right-mouse-button to display the option menu.

5. Select **Continue** in the popup menu to proceed with the creation of the balloon.

6. On your own, repeat the above steps and complete the drawing as shown.

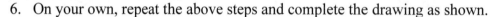
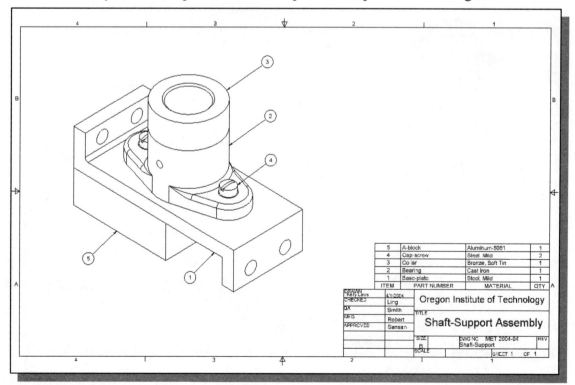

5	A-block	Aluminum-6061	1
4	Cap-screw	Steel Mild	2
3	Collar	Bronze, Soft Tin	1
2	Bearing	Cast Iron	1
1	Base-plate	Steel, Mild	1
ITEM	PART NUMBER	MATERIAL	QTY

Oregon Institute of Technology

Shaft-Support Assembly

Bill of Materials

❖ A bill of materials (BOM) is a table that contains information about the parts within an assembly. The BOM can include information such as part names, quantities, costs, vendors, and all of the other information related to building the part. The *parts list*, which is used in an assembly drawing, is usually a partial list of the associated BOM.

In *Autodesk Inventor*, both the *bill of materials* and *parts list* can be derived directly from data generated by the assembly and the part properties. We can select which properties to be included in the *bill of materials* or *parts list*, in what order the information is presented, and in what format to export the information. The exported file can be used in an application such as a spreadsheet or text editor.

(a) BOM from Parts List

1. Move the cursor on top of the *Parts List* and right-mouse-click once to display the option menu.

2. Choose **Export...** in the option menu as shown.

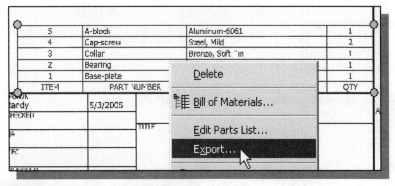

3. Click on the *Save as type* list and choose **Microsoft Excel** as shown.

4. Enter *Shaft-support.xls* as the filename and click **Save** to export the BOM.

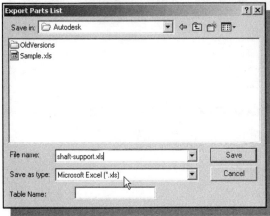

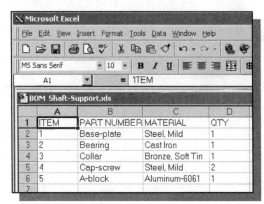

❖ On your own, examine the exported BOM by opening up the file in Excel.

(b) BOM from Assembly Model

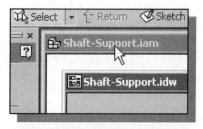

1. Click on the *Shaft-Support.iam* window to switch back to the assembly model.

2. Click **Tools** in the pull-down menu.

3. Select **Bill of Materials** in the options list.

❖ Note that many of the controls and options are similar to those of the **Parts List** command in the *Drawing Mode*.

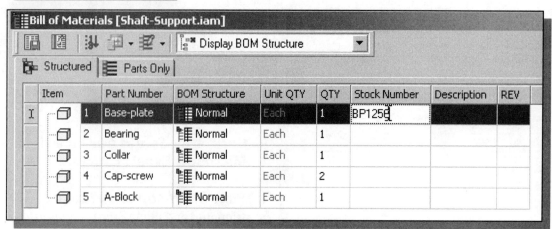

4. Click inside the *Stock Number* box to enter the *Edit Mode*.

5. Enter **BP1256** as the new *Stock Number*.

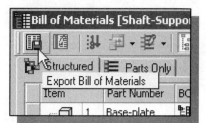

6. Click the **Export Bill of Materials** button.

7. On your own, export using the Microsoft Excel format and examine the *BOM* in Microsoft Excel.

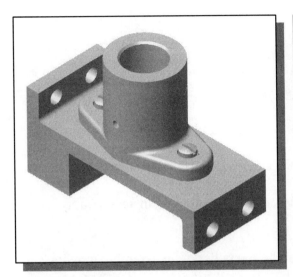 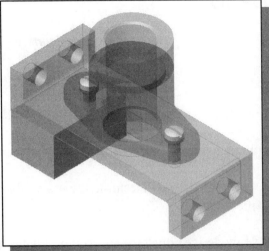

Conclusion

Design includes all activities involved from the original concept to the finished product. Design is the process by which products are created and modified. For many years designers sought ways to describe and analyze three-dimensional designs without building physical models. With advancements in computer technology, the creation of parametric models on computers offers a wide range of benefits. Parametric models are easier to interpret and can be easily altered. Parametric models can be analyzed using finite element analysis software, and simulation of real-life loads can be applied to the models and the results graphically displayed.

Throughout this text, various modeling techniques have been presented. Mastering these techniques will enable you to create intelligent and flexible solid models. The goal is to make use of the tools provided by *Autodesk Inventor* and to successfully capture the **DESIGN INTENT** of the product. In many instances, only a single approach to the modeling tasks was presented; you are encouraged to repeat all of the lessons and develop different ways of thinking in accomplishing the same tasks. We have only scratched the surface of *Autodesk Inventor*'s functionality. The more time you spend using the system, the easier it will be to perform parametric modeling with *Autodesk Inventor*.

Summary of Modeling Considerations

- **Design Intent** – determine the functionality of the design; select features that are central to the design.

- **Order of Features** – consider the parent/child relationships necessary for all features.

- **Dimensional and Geometric Constraints** – the way in which the constraints are applied determines how the components are updated.

- **Relations** – consider the orientation and parametric relationships required between features and in an assembly.

Questions:

1. What is the purpose of using *assembly constraints*?

2. List three of the commonly used *assembly constraints*.

3. Describe the difference between the **Mate** constraint and the **Flush** constraint.

4. In an assembly, can we place more than one copy of a part? How is it done?

5. How should we determine the assembly order of different parts in an assembly model?

6. How do we adjust the information listed in the **parts list** of an assembly drawing?

7. In *Autodesk Inventor*, describe the procedure to create a **bill of materials** (BOM)?

8. Create sketches showing the steps you plan to use to create the four parts required for the assembly shown on the next page:

Ex.1)

Ex.2)

Ex.3)

Ex.4)

Exercises:

1. **Wheel Assembly** (Create a set of detail and assembly drawings. All dimensions are in mm.)

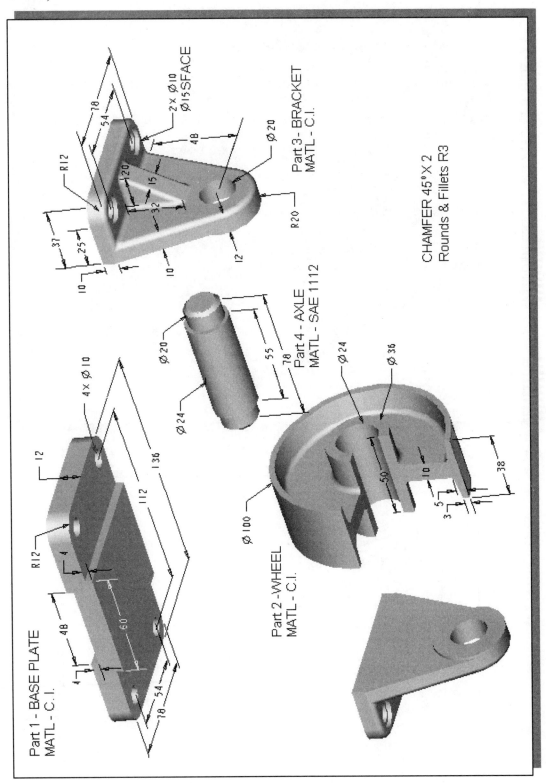

2. **Leveling Assembly** (Create a set of detail and assembly drawings. All dimensions are in mm.)

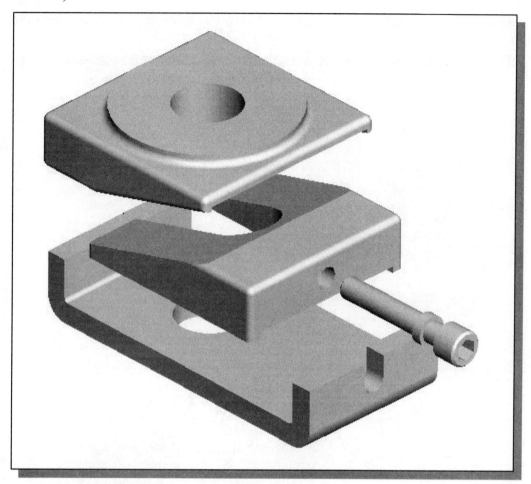

(a) **Base Plate**

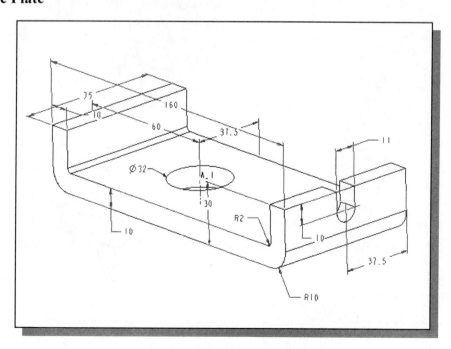

(b) **Sliding Block** (Rounds & Fillets: R3)

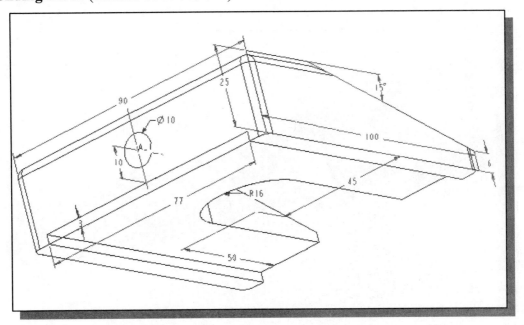

(c) **Lifting Block** (Rounds & Fillets: R3)

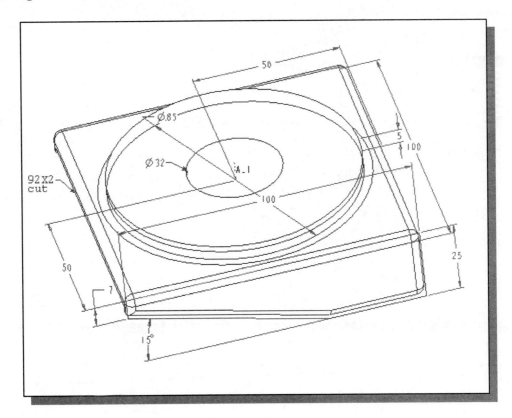

(d) Adjusting Screw (M10 × 1.5)

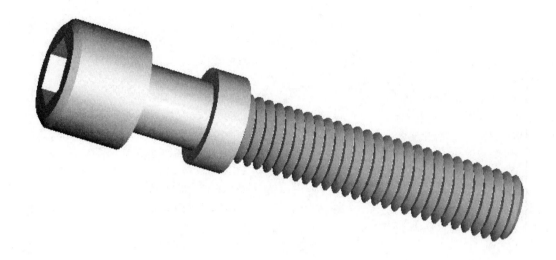

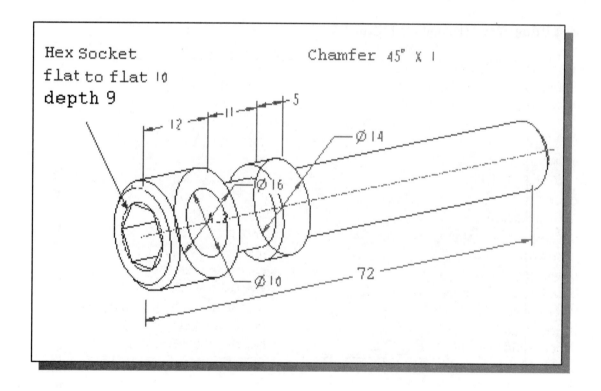

Hex Socket
flat to flat 10
depth 9

Chamfer 45° X 1

12

11

5

Ø 14

Ø 16

Ø 10

72

INDEX

A

Accept Icon, 2-11
Activate, 8-9
Active Constraint Box, 5-5
Adaptive, 12-27
Adaptive Design approach, 12-2, 12-25
Alignment, Break, 8-14
Angle Constraint, 12-11, 12-12
Angled Work Plane, 9-8
Annotation Toolbar, 8-14
ANSI-A, Title Block, 9-18
ANSI Standard, 8-5
Application Options, 7-4
Arc, Center point arc, 7-13
Arrowheads Inside, 8-17
Assembly Modeling Methodology, 12-3
Assembly constraints, 12-11
Associative Functionality, 8-2, 8-23
Autodesk Inventor, Startup, 1-7
Auto Dimension, 5-12, 6-16
Automatically Retrieve Dimensions, 10-23
Auxiliary Views, 9-2, 9-19

B

Balloon, 12-39
Base component, 12-8
Base Feature, 3-8
Base View, 9-20
Base Orphan Reference Node, 7-2
Bill of materials (BOM), 12-39
Binary Tree, 3-2
Block, solids, 3-2
BOM, 12-39
Boolean Intersect, 3-2
Boolean Operations, 3-2
BORN, 7-2
Border, new, 10-20
Bottom Up Approach, 12-3
Boundary representation (B-rep), 1-5
Browser window, 1-11, 4-2, 4-5
Buttons, Mouse, 1-12

C

CAE, 1-2
Canceling commands, 1-12
Cartesian coordinate system, 2-18
Center Line Bisector, 8-19
Center Mark, 8-19
Center option, 4-14
Center point arc, 7-13
Center-point Circle, 2-24
Centered Pattern, 10-30
Child, 7-2
Choose Template, 3-6
CIRCLE, Center point, 2-24
Circular Pattern, 10-17
Click and drag, 6-7
Coil and Thread, 12-6
Coincident Constraint, 2-7, 5-10, 5-11
Colinear Constraint, 5-6
Command Toolbar, 1-9
Computer Geometric Modeling, 1-2
Concentric constraint, 2-7, 5-6
Constrained move, 12-16
Constraint, Delete, 5-11
Constraints, Geometric, 2-7, 5-2, 5-6
Constraints Settings, 5-16
Constraint, Show, 5-4
Constraints Box, 5-5
Constraints Box, Pin, 5-5
Construction geometry, 10-12
Construction lines, 10-14
Constructive solid geometry, 3-2
Coordinate systems, 2-18
Create Component, 12-25
Create Feature, Extrude, 12-26
CSG, 1-5, 3-2
Cut, All, 2-26
Cut, Boolean, 3-2
Cut, Extrude, 2-26
Cutting Plane Line, 8-13

NOTES:

NOTES:

NOTES:

NOTES:

NOTES:

NOTES:

NOTES:

NOTES:

NOTES:

NOTES:

NOTES:

NOTES:

NOTES: